EPA 600/R-13/174 | August 2013
www.epa.gov/ord

United States
Environmental Protection
Agency

Watershed Management Optimization Support Tool (WMOST) v1

USER MANUAL
AND CASE STUDY EXAMPLES

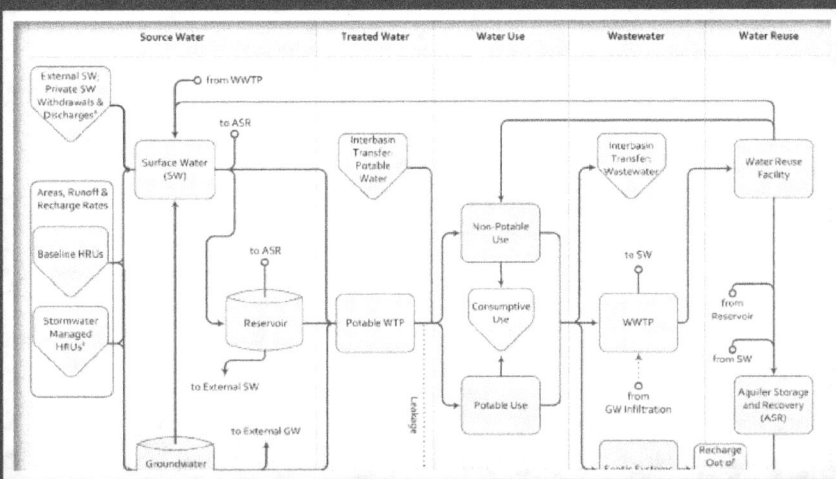

Office of Research and Development
National Health and Environmental Effects Research Laboratory, Atlantic Ecology Division

EPA 600/R-13/174 | August 2013

Watershed Management Optimization Support Tool (WMOST) v1

User Manual and Case Study Examples

EPA Project Team

Naomi Detenbeck and Marilyn ten Brink,
NHEERL, Atlantic Ecology Division
Narragansett, RI 02882

Alisa Morrison
Student Services Contractor at ORD, NHEERL, Atlantic Ecology Division
Narragansett, RI 02882

Yusuf Mohamoud
ORD, NERL, Ecosystems Research Division
Athens, GA 30605

Ralph Abele and Jackie LeClair
Region 1
Boston, MA 02109

Abt Associates Project Team

Viktoria Zoltay, Becky Wildner, Lauren Parker and Isabelle Morin, Abt Associates, Inc.
Nigel Pickering, Horsley Witten Group under subcontract to Abt Associates Inc.
Richard M. Vogel, Tufts University under subcontract to Abt Associates Inc

Notice

The information in this document has been funded by the U.S. Environmental Protection Agency (EPA), in part by EPA's Green Infrastructure Initiative, under EPA Contract No. EP-C-07-023/ Work Assignment 32 to Abt Associates, Inc. It has been subjected to the Agency's peer and administrative review, and it has been approved for publication. Mention of trade names or commercial products does not constitute endorsement or recommendation for use.

Although a reasonable effort has been made to assure that the results obtained are correct, the computer programs described in this manual are experimental. Therefore, the author and the U.S. Environmental Protection Agency are not responsible and assume no liability whatsoever for any results or any use made of the results obtained from these programs, nor for any damages or litigation that result from the use of these programs for any purpose.

Abstract

The Watershed Management Optimization Support Tool (WMOST) is intended to be used as a screening tool as part of an integrated watershed management process such as that described in EPA's watershed planning handbook (EPA 2008).[1] The objective of WMOST is to serve as a public-domain, efficient, and user-friendly tool for local water resources managers and planners to screen a wide-range of potential water resources management options across their watershed or jurisdiction for cost-effectiveness as well as environmental and economic sustainability (Zoltay et al 2010). Examples of options that could be evaluated with the tool include projects related to stormwater, water supply, wastewater and water-related resources such as Low-Impact Development (LID) and land conservation. The tool is intended to aid in evaluating the environmental and economic costs, benefits, trade-offs and co-benefits of various management options. In addition, the tool is intended to facilitate the evaluation of low impact development (LID) and green infrastructure as alternative or complementary management options in projects proposed for State Revolving Funds (SRF). WMOST is a screening model that is spatially lumped with a daily or monthly time step. The model considers water flows but does not yet consider water quality. The optimization of management options is solved using linear programming. The target user group for WMOST consists of local water resources managers, including municipal water works superintendents and their consultants. This document includes a user guide and presentation of two case studies as examples of how to apply WMOST. Theoretical documentation is provided in a separate report (EPA/600/R-13/151).

Keywords: Integrated watershed management, water resources, decision support, optimization, green infrastructure

---

[1] EPA. 2008. Handbook for Developing Watershed Plans to Restore and Protect Our Waters. March 2008. US Environmental Protection Agency. Office of Water. Nonpoint Source Control Branch, Washington, D.C. EPA 841-B-08-002

Preface

Integrated Water Resources Management (IWRM) has been endorsed for use at multiple scales. The Global Water Partnership defines IWRM as "a process which promotes the coordinated development and management of water, land and related resources, in order to maximize the resultant economic and social welfare in an equitable manner without compromising the sustainability of vital ecosystems".[2] IWRM has been promoted as an integral part of the "Water Utility of the Future"[3] in the United States. The American Water Resources Association (AWRA) has issued a position statement calling for implementation of IWRM across the United States and committed the AWRA to help strengthen and refine IWRM concepts.[4] The U.S. Environmental Protection Agency (EPA) has also endorsed the concept of IWRM, focusing on coordinated implementation of stormwater and wastewater management.[5]

Several states and river basin commissions have started to implement IWRM.[6] Even in EPA's Region 1 where water is relatively plentiful, states face the challenge of developing balanced approaches for equitable and predictable distribution of water resources to meet both human and aquatic life needs during seasonal low flow periods and droughts. The state of Massachusetts recently spearheaded the Sustainable Water Management Initiative (SWMI) process to allocate water among competing human and aquatic life uses in a consistent and sustainable fashion.[7]

Stormwater and land use management are two aspects of IWRM which include practices such as green infrastructure (GI, both natural GI and constructed stormwater BMPs), low-impact development (LID) and land conservation. In recent years, the EPA's SRF funding guidelines have been broadened to include support for green infrastructure at local scales–e.g., stormwater best management practices (BMPs) to reduce runoff and increase infiltration–and watershed scales–e.g., conservation planning for source water protection. Despite this development, few applicants have

---

[2] UNEP-DHI Centre for Water and Environment. 2009. Integrated Water Resources Management in Action. WWAP, DHI Water Policy, UNEP-DHI Centre for Water and Environment.

[3] NACWA, WERF, and WEF. 2013. The Water Resources Utility of the Future: A Blueprint for Action. National Association of Clean Water Agencies (NACWA), Water Environment Research Foundation (WERF) and Water Environment Federation (WEF), Washington, D.C.

[4] http://www.awra.org/policy/policy-statements--water-vision.html

[5] Nancy Stoner memo: http://water.epa.gov/infrastructure/greeninfrastructure/upload/memointegratedmunicipalplans.pdf

[6] AWRA. 2012. Case Studies in Integrated Water Resources Management: From Local Stewardship to National Vision. American Water Resources Association Policy Committee, Middleburg, VA.

[7] MA EAA. 2012. Massachusetts Sustainable Water Management Initiative Framework Summary (November 28, 2012); http://www.mass.gov/eea/agencies/massdep/water/watersheds/sustainable-water-management-initiative-swmi.html

---

taken advantage of these opportunities to try nontraditional approaches to water quality improvement.[8] In a few notable cases, local managers have evaluated the relative cost and benefit of preserving green infrastructure compared to traditional approaches. In those cases, the managers have championed the use of green infrastructure as part of a sustainable solution for IWRM but these examples are rare.[9]

Beginning with the American Recovery and Reinvestment Act (ARRA), and continued with 2010 Appropriations language, Congress mandated a 20% set-aside of SRF funding for a "Green Project Reserve (GPR)", which includes green infrastructure and land conservation measures as eligible projects in meeting water quality goals. The utilization of the GPR for green infrastructure projects has been relatively limited and responses have varied widely across states. According to a survey of 19 state allocations of Green Project Reserve funds, only 18% of funds were dedicated to green infrastructure projects and none of these projects were categorized as conservation planning to promote source water protection.[7] The state of Virginia passed regulations banning the use of ARRA funds for green infrastructure projects until after wastewater treatment projects had been funded.[7] In New England, states exceeded the 20% GPR mandate and used 30% of their ARRA funds for the GPR, but directed most of the funds (76%) to energy efficiency and renewables; other uses of ARRA funds included 12% for water efficiency, 9% for green infrastructure, and 3% for environmentally innovative projects.

In order to assist communities in the evaluation of GI, LID, and land conservation practices as part of an IWRM approach, EPA Office of Research and Development, in partnership with EPA Region 1, supported the development of the Watershed Management Optimization Support Tool (WMOST). WMOST is based on a recent integrated watershed management optimization model that was created to allow water resources managers to evaluate a broad range of technical, economic, and policy management options within a watershed.[10] This model includes evaluation of conservation options for source water protection and infiltration of stormwater on forest lands, green infrastructure stormwater BMPs to increase infiltration, and other water-related management options. The current version of

---

[8] American Rivers. 2010. Putting Green to Work: Economic Recovery Investments for Clean and Reliable Water. American Rivers, Washington, D.C

[9] http://www.crwa.org/blue.html, http://v3 mmsd.com/greenseamsvideo1.aspx

[10] Zoltay, V.I. 2007. Integrated watershed management modeling: Optimal decision making for natural and human components. M.S. Thesis, Tufts Univ., Medford, MA.; Zoltay, V.I., R.M. Vogel, P.H. Kirshen, and K.S. Westphal. 2010. Integrated watershed management modeling: Generic optimization model applied to the Ipswich River Basin. Journal of Water Resources Planning and Management.

WMOST focuses on management options for water quantity endpoints. Additional functionality to address water quality issues is one of the high priority enhancements identified for future versions.

Development of the WMOST tool was overseen by an EPA Planning Team. Priorities for update and refinement of the original model[9] were established following review by a Technical Advisory Group comprised of water resource managers and modelers. Case studies for each of three communities were developed to illustrate the application of IWRM using WMOST; two of these case studies (Upper Ipswich River, and Danvers/Middleton, MA) are presented here. WMOST was presented to stakeholders in a workshop held at the EPA Region 1 Laboratory in Chelmsford, MA in April 2013, with a follow-up webinar on the Danvers/Middleton case study in May 2013. Feedback from the Technical Advisory Group and workshop participants has been incorporated into the user guide and theoretical documentation for WMOST.

Acknowledgements

WMOST builds on research funded by the National Science Foundation Graduate Research Fellowship Program and published in Zoltay, V. Kirshen, P.H. Vogel, R.M. and Westphal, K.S. 2010. "Integrated Watershed Management Modeling: Optimal Decision Making for Natural and Human Components." Journal of Water Resources Planning and Management, 136:5, 566-575.

EPA Project Team
Naomi Detenbeck and Marilyn ten Brink, U.S. EPA ORD, NHEERL, Atlantic Ecology Division
Alisa Morrison, Student Services Contractor at U.S. EPA ORD, NHEERL, Atlantic Ecology Division
Ralph Abele and Jackie LeClair, U.S. EPA Region 1
Yusuf Mohamoud, U.S. EPA ORD, NERL, Ecosystems Research Division

Technical Advisory Group
Alan Cathcart, Concord, MA Water/Sewer Division
Greg Krom, Topsfield, MA Water Department
Dave Sharples, Somersworth, NH Planning and Community Development
Mark Clark, North Reading, MA Water Department
Peter Weiskel, U.S. Geological Survey, MA-RI Water Science Center
Kathy Baskin, Massachusetts Executive Office of Energy and Environmental Affairs
Steven Estes-Smargiassi, Massachusetts Water Resources Authority
Hale Thurston, U.S. EPA ORD, NRMRL, Sustainable Technology Division
Rosemary Monahan, U.S. EPA Region 1
Scott Horsley, Horsley Witten Group
Kirk Westphal, CDM Smith
James Limbrunner, Hydrologics, Inc.
Jay Lund, University of California, Davis

Reviewers
Theoretical Documentation
Marisa Mazzotta, U.S. EPA ORD, NHEERL, Atlantic Ecology Division
Mark Voorhees, U.S. EPA Region 1
Michael Tryby, U.S. EPA ORD, NERL, Ecosystems Research Division
WMOST Tool, User Guide and Case Studies
Daniel Campbell, U.S. EPA ORD, NHEERL, Atlantic Ecology Division
Alisa Richardson, Rhode Island Department of Environmental Management (partial review)
Alisa Morrison, Student Services Contractor at U.S. EPA ORD, NHEERL, Atlantic Ecology Division
Jason Berner, U.S. EPA OW, OST, Engineering Analysis Division

Table of Contents

Exhibits

1. Background

1.1 Objective of the Tool

The Watershed Management Optimization Support Tool (WMOST) is a public-domain software application designed to aid decision making in integrated water resources management. WMOST is intended to serve as an efficient and user-friendly tool for water resources managers and planners to screen a wide-range of strategies and management practices for cost-effectiveness and environmental sustainability in meeting watershed or jurisdiction management goals (Zoltay et al 2010).[10]

WMOST identifies the least-cost combination of management practices to meet the user specified management goals. Management goals may include meeting projected water supply demand and minimum and maximum in-stream flow targets. The tool considers a range of management practices related to water supply, wastewater, nonpotable water reuse, aquifer storage and recharge, stormwater, low-impact development (LID) and land conservation, accounting for the both the cost and performance of each practice. In addition, WMOST may be run for a range of values for management goals to perform a cost-benefit analysis and obtain a Pareto frontier or trade-off curve. For example, running the model for a range of minimum in-stream flow standards provides data to create a trade-off curve between increasing in-stream flow and total annual management cost.

WMOST is intended to be used as a screening tool as *part* of an integrated watershed management process such as that described in EPA's watershed planning handbook (EPA 2008),[1] to identify the strategies and practices that seem most promising for more detailed evaluation. For example, results may demonstrate the potential cost-savings of coordinating or integrating the management of water supply, wastewater and stormwater. In addition, the tool may facilitate the evaluation of LID and green infrastructure as alternative or complementary management options in projects proposed for State Revolving Funds (SRF). As of October 2010, SRF Sustainability Policy calls for integrated planning in the use of SRF resources as a means of improving the sustainability of infrastructure projects and the communities they serve. In addition, Congress mandated a 20% set-aside of SRF funding for a "Green Project Reserve" which includes green infrastructure and land conservation measures as eligible projects in meeting water quality goals.

1.2 Overview

WMOST combines an optimization framework with water resources modeling to evaluate the effects of management decisions within a watershed context. The watershed system modeled in WMOST version 1 is shown in *Exhibit 1*. The exhibit shows the *possible* watershed system components and *potential* water flows among them.

Exhibit 1. Schematic of Potential Water Flows in the WMOST

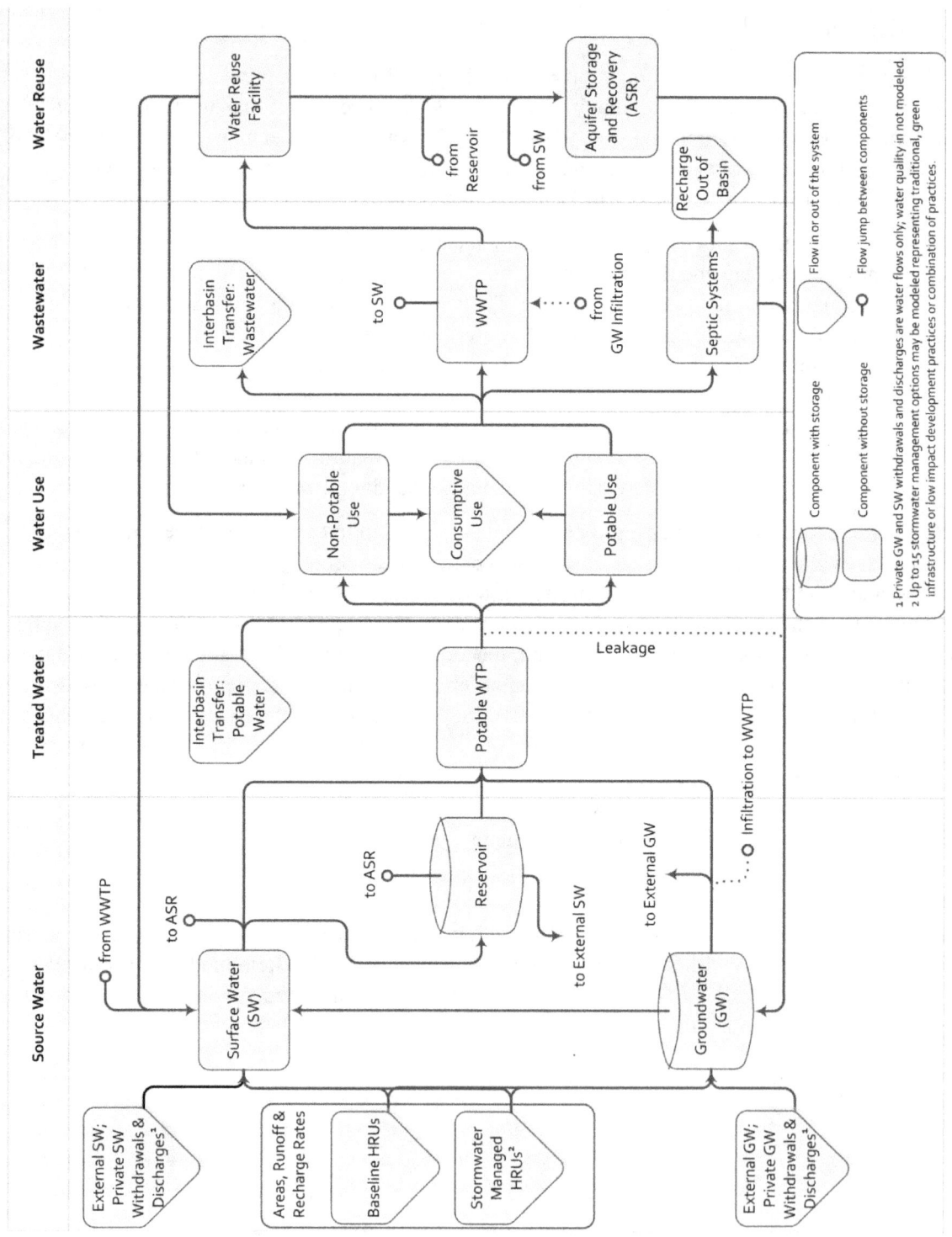

The principal characteristics of WMOST include:

- Implementation in Microsoft Excel 2010© which is linked seamlessly with Visual Basic for Applications (VBA) and a free, linear programming (LP) optimization solver, eliminating the need for specialized software and using the familiar Excel platform for the user interface;
- User-specified inputs for characterizing the watershed, management practices, and management goals and generating a customized optimization model (see *Exhibit 2* for a list of available management practices and goals);
- Use of Lp_solve 5.5, a LP optimization solver, to determine the least-cost combination of practices that achieves the user-specified management goals (See *Section 3* in the separate Theoretical Documentation for details on Lp_solve 5.5, LP optimization, and the software configuration);
- Spatially lumped calculations modeling one basin and one reach but with flexibility in the number of hydrologic response units (HRUs),[11] each with an individual runoff and recharge rate;
- Modeling time step of a day or month without a limit on the length of the modeling period;[12]
- Solutions that account for both the direct and indirect effects of management practices (e.g., since optimization is performed within the watershed system context, the model will account for the fact 1) that implementing water conservation will reduce water revenue, wastewater flow and wastewater revenue if wastewater revenue is calculated based on water flow or 2) that implementing infiltration-based stormwater management practices will increase aquifer recharge and baseflow for the stream reach which can help meet minimum in-stream flow requirements during low precipitation periods, maximum in-stream flow requirements during intense precipitation seasons, and water supply demand from increased groundwater supply);
- Ability to specify up to fifteen stormwater management options, including traditional, green infrastructure or LID practices;
- A sustainability constraint that forces the groundwater and reservoir volumes at the start and end of the modeling period to be equal;
- Enforcement of physical constraints, such as the conservation of mass (i.e., water), within the watershed; and
- Consideration of water flows only (i.e., no water quality modeling yet).

The rest of this document is organized as follows. *Section 2* provides directions for basic model setup and application with screenshots as well as the steps for performing sensitivity and trade-off analyses. A case study example for a watershed is presented in *Section 3.1* for the Upper Ipswich River Watershed. Another case study for two towns sharing one water utility is presented in *Section 3.2* for Danvers and Middleton, Massachusetts. The WMOST files for these case studies for all scenarios are available and may be used as a source of default data, especially for similar watersheds and similar sized water and wastewater systems.

---

[11] Land cover, land use, soil, slope and other land characteristics affect the fraction of precipitation that will runoff, recharge and evapotranspire. Areas with similar land characteristics that respond similarly to precipitation are termed hydrologic response units.

[12] While the number of HRUs and modeling period are not limited, solution times are significantly affected by these model specifications.

A *separate* Theoretical Documentation report provides a detailed description of WMOST including a mathematical description and the internal configuration of the software applications that constitute the model.

Exhibit 2. Summary of Management Goals and Management Practices[13]

Management Practice	Action[14]	Model Component Affected	Impact
Land conservation	Increase area of land use type specified as 'conservable'	Land area allocation	Preserve runoff & recharge quantity & quality
Stormwater management via traditional, green infrastructure or low impact development practices	Increase area of land use type treated by specified management practice	Land area allocation	Reduce runoff, increase recharge, treatment
Surface water storage capacity	Increase maximum storage volume	Reservoir/Surface Storage	Increase storage, reduce demand from other sources
Surface water pumping capacity	Increase maximum pumping capacity	Potable water treatment plant	Reduce quantity and/or timing of demand from other sources
Groundwater pumping capacity	Increase maximum pumping capacity	Potable water treatment plant	Reduce quantity and/or timing of demand from other sources
Change in quantity of surface versus groundwater pumping	Change in pumping time series for surface and groundwater sources	Potable water treatment plant	Change the timing of withdrawal impact on water source(s)
Potable water treatment capacity	Increase maximum treatment capacity	Potable water treatment plant	Treatment to standards, meet potable human demand
Leak repair in potable distribution system	Decrease % of leaks	Potable water treatment plant	Reduce demand for water quantity
Wastewater treatment capacity	Increase million gallons per day (MGD) treated	Wastewater treatment plant	Maintain water quality of receiving water (or improve if sewer overflow events)
Infiltration repair in wastewater collection system	Decrease % of leaks	Wastewater treatment plant	Reduce demand for wastewater treatment capacity

---

[13] The user may specify which practices are available for their study area and are to be included in the optimization. Directions for this are provided with each practice in the User Manual and WMOST interface.

[14] Please refer to the separate Theoretical Documentation for the specific effect of each management practice.

Management Practice	Action[14]	Model Component Affected	Impact
Water reuse facility (advanced treatment) capacity	Increase MGD treated	Water reuse facility	Produce water for nonpotable demand, Aquifer storage & recharge (ASR), and/or improve water quality of receiving water
Nonpotable distribution system	Increase MGD delivered	Nonpotable water use	Reduce demand for potable water
Aquifer storage & recharge (ASR) facility capacity	Increase MGD treated & injected	ASR facility	Increase recharge, treatment, and/or supply
Demand management by price increase	Increase % of price	Potable and nonpotable water and wastewater	Reduce demand
Direct demand management	Percent decrease in MGD	Potable and nonpotable water and wastewater	Reduce demand
Interbasin transfer – potable water import capacity	Increase or decrease MGD delivered	Interbasin transfer – potable water import	Increase potable water supply or reduce reliance on out of basin sources
Interbasin transfer – wastewater export capacity	Increase or decrease MGD delivered	Interbasin transfer – wastewater export	Reduce need for wastewater treatment plant capacity or reduce reliance on out of basin services
Minimum human water demand	Minimum MGD	Groundwater and surface water pumping and/or interbasin transfer	Meet human water needs
Minimum in-stream flow	Minimum ft^3/sec	Surface water	Meet in-stream flow standards, improve ecosystem health and services, improve recreational opportunities
Maximum in-stream flow	Maximum ft^3/sec	Surface water	Meet in-stream flow standards, improve ecosystem health and services by reducing scouring, channel and habitat degradation, and decrease loss of public and private assets due to flooding

2. Getting Started

WMOST is a screening tool for watershed management and planning. One of the envisioned applications of WMOST is determining the least cost combination of management options to meet management goals for a town or watershed's planning horizon. For example, the water works portion of a town's master plan may ask, "What stormwater practices must be installed, demand management programs created and/or infrastructure capacity constructed to meet projected human demand for the next 20 years while meeting minimum and maximum in-stream flow targets to preserve stream health?" To address such a planning question, all input data must correspond to the conditions projected to occur by the end of a 20 year planning period. For example, human demand would need to be projected 20 years from the planning year. Most of the User Guide is written from the perspective of a user who is screening management practices to address such planning questions and suggestions are provided throughout the User Guide and in the case study appendices for how to specify input data appropriately. As such, the model does not provide an annual implementation plan or specifics on operations of systems. Rather it provides the management practices and associated costs that meet management goals at least cost and the state of the watershed and human system at the end of the planning period if the management practices have been implemented.

2.1 Preparing for a Model Run

This section describes model specifications the user must consider prior to applying WMOST. All data sources for the case studies are detailed in the appendices. Some of those data sources, especially for environmental data, are state or national level and may serve as a source for your project. Most data related to the human water system is anticipated to be available to the municipality(ies) from their own internal sources.

Defining Hydrologic Response Units

A main input data requirement is runoff and recharge rates (RRRs) for hydrologic response units (HRUs)[15] within the study area and the corresponding area for each HRU. These data may be derived from a calibrated/validated simulation model such as Hydrological Simulation Program Fortran (HSPF),[16] Soil Water and Assessment Tool (SWAT)[17] and/or Storm Water Management Model.[18] If a watershed simulation model is not available for the study area (e.g., from U.S. Geological Survey) and resources do not allow for the creation and setup of a model, then the user may try using default rates from models run for watersheds with similar characteristics. Additionally there may be generic RRRs available from state or regional studies. Such rates would specify the HRU characteristics for which the rates are applicable. A geographic information system can then be used to determine the area associated with each HRU in you study area. Future versions of the model may include default RRRs for HRUs for various watersheds and/or ecoregions.

---

[15] Land cover, land use, soil, slope and other land characteristics affect the fraction of precipitation that will runoff, recharge and evapotranspire. Areas with similar characteristics – hydrologic response units (HRUs)[15] – respond similarly to precipitation.

[16] http://water.usgs.gov/software/HSPF/

[17] http://swat.tamu.edu/

[18] http://www.epa.gov/nrmrl/wswrd/wq/models/swmm/

In addition to a baseline set of HRUs, up to 15 "sets" of "managed" HRUs may be specified with corresponding areas, RRRs and management costs. For HRU sets, the baseline set is used to specify runoff and recharge rates and areas for HRUs for the baseline conditions of the model run. For managed sets, you may specify runoff and recharge rates that reflect some form of land management practice and the associated cost. These managed RRRs may be derived using SWMM or other stormwater management models.

For urban HRUs, the "managed set" may reflect RRRs resulting from use of a stormwater best management practice (e.g., bioretention basin, swales) and/or low impact development with reduced impervious area. For agricultural HRUs, the "managed set" may reflect RRRs resulting from implementation of edge of site or riparian buffers. For each set, you can specify the area of each HRU on which the management practice may be implemented. Therefore, for the stormwater managed set, you may restrict available area to urban HRUs only and vice-versa for agricultural management. In addition, if stormwater management exists in part of the watershed, urban HRUs may be defined separately for areas that are under management and areas that are not under management with their respective RRRs. Then, under managed HRU sets, the new stormwater management practice may be limited to unmanaged, urban HRUs and excluded for managed, urban HRUs (as well as other HRUs such as agricultural or forest).

Defining the Study Area

Ideally, the study area is the entire land area draining to the stream reach of interest; however, jurisdictional boundaries often cut across subbasins. This requires that the hydrology is modeled at the subbasin or watershed level while management practices are limited to those areas within the jurisdiction(s) cooperating in the management plan. The second case study of Danvers and Middleton, MA shows the example of how to use the model in such circumstances. The first case study of the Upper Ipswich River Basin assumes that the entire watershed is cooperating in the management strategy such as in a water district and, therefore, management practices are specified to be available for the entire watershed.[19]

Defining the Modeling Time Period

The model may be run on a daily or monthly time step. The user may choose the time step depending on the temporal resolution of available input data, desired management practices and/or known system behavior. For example, if stormwater management practices will be considered, a daily time step is advised as storm events and their effects are observable on a time scale closer to a daily rather than monthly time step. If the user desires to know the monthly or approximate water balance for watershed or human system components, then a monthly time step would be sufficient.

The user should run the model for multiple years that cover dry, average and wet years of precipitation. That is, time series that are input (e.g., RRRs, human demand, surface water inflow from upstream) should include a range of potential conditions. Ensuring these specifications are met will ensure that the management solution screened by the model will be sustainable over a range of

---

[19] If the user wants to model multiple adjacent/downstream study areas, theoretically, the time series of surface water outflow from the upstream study area may be used an input into the downstream study area. WMOST v1 does not output this time series in table form (only as a graph) but this functionality is listed for future development. In addition, enhanced spatial modeling is identified as an area for future development so that all areas or reaches can be optimized simultaneously rather than just consecutively from upstream to downstream reaches.

potential future conditions. In addition, the user is advised to run not only a range of historical conditions but future, projected conditions. This may be accomplished, for example, by adjusting historical conditions for projected climate change. The EPA website "Climate Change Impacts and Adapting to Change" describes projected changes by region.[20] EPA's Climate Resilience Evaluation and Awareness Tool provides projected changes in temperature and precipitation for climate stations throughout the United States.[21] These values may be used to adjust the detailed watershed simulation model from which watershed time series data is obtained for WMOST (e.g., see Soil and Water Assessment Tool climate change function) and to adjust the traditional methodology used for projecting human demand.

Note that running WMOST with RRRs and other environmental data from a specific time period such as 2005-2010 does not necessarily represent watershed conditions that only occurred during those years but watershed conditions that would occur in a similar 5-year period of weather. Therefore, these data can be adjusted for climate change or other uncertainties and re-run to determine the sensitivity of the solution, that is, combination of management practices and costs, to potential future deviations from historical conditions. In fact, **the user is highly encouraged to perform sensitivity analyses especially on input data with least certainty to determine the robustness of the solution**. *Section 2.4* briefly describes the process for performing sensitivity analyses.

2.2   Setting Up and Running the Model

System Requirements

To open and run WMOST, you will need Microsoft Excel version 2010 installed on your computer. The WMOST Excel file and the file for the solver, *lpsolve55.dll*, must be placed in the same folder. After opening WMOST, choose 'Enable content' or 'Enable macros' if these prompts are displayed.

When using WMOST, you may save various versions that are set up for different scenarios. You cannot run multiple scenarios at the same time from the same folder. However, you may save a different scenario along with the *lpsolve55.dll* file in a different folder in order to run multiple scenarios at once. Depending on your computer's specifications this may increase the run time for each model.

If other Excel files are open while running WMOST, the Results table will have the correct values but may not be formatted properly. Therefore, it is recommended that you do not have other Excel files open and run model scenarios one at a time.

Finally, if you encounter software errors, please email Naomi Detenbeck at detenbeck.naomi@epa.gov with the subject "WMOST bug". To register for notices of patches and new releases, please email the same address with the subject line "WMOST register".

The User Interface–Step by Step

When you open WMOST you will see the familiar Excel interface with one worksheet called "Main". You can navigate to input tables using the blue buttons and result table and figures using the green buttons found on this screen. To begin entering data for your study area, start by completing input fields on the "Main" worksheet. All input fields are blue boxes.

---

[20] http://www.epa.gov/climatechange/impacts-adaptation/

[21] http://water.epa.gov/infrastructure/watersecurity/climate/creat.cfm

Please note that example screenshots and values displayed in them are from the Danvers-Middleton case study and are not necessarily appropriate values for your study area. WMOST performs several basic checks to ensure that input data requirements are met, for example, that price elasticies are negative and minimum in-stream flow targets are smaller than maximum in-stream flow targets. If these basic requirements are not met, the user is informed with a message box and asked to re-enter the information. *Section 6.1* in the Theoretical Documentation provides additional details on input data checks and user support.

Step 1. HRUs, Areas, Runoff and Recharge. Enter the number of HRU types and HRU sets that you intend to model. HRUs are areas of similar hydrology based on similar characteristics such as land use, soil and/or slope. The number of HRUs will likely be determined by the diversity of these land characteristics in your study area and your source of runoff and recharge rates. For example, a detailed simulation model that may be available for your study area may have predefined HRUs.

INPUT DATA

1. Enter the number of HRU types in your study area and the number of land management options you will model.

Number of HRU Types: (11)   Number of HRU Sets (baseline plus managed sets): (7)

Step 2. Press the "Setup 1" button to automatically prepare input tables for land use, runoff and recharge data based on your values from Step 1. *The process creates blank input tables; therefore, do not press this button again unless you have your input data saved elsewhere and want to change the number of HRUs or HRU sets.*

2. Press "Setup 1" button to prepare input tables for land use, runoff, and recharge data.

Step 3. Here, the "Land Use", "Runoff" and "Recharge" buttons will direct you to their respective input tables:

Selecting the "Land Use" button directs you to the land-use input screen shown below.

Baseline HRUs and Their Limits with Respect to Land Conservation

HRU ID	HRU Name	Baseline Area [acre]	Minimum Area [acre]	Maximum Area [acre]	Initial Cost to Conserve [$/acre]	O&M Cost [$/acre/yr]
HRU1B	Forest, sand & gravel	1,681	1,681	2,760	187,408	1,874
HRU2B	Open, sand & gravel	437	437	774	187,408	1,874
HRU3B	Low-resid, sand & gravel	3,099	364	3,099	-9	-9
HRU4B	High-resid, sand & gravel	1,274	1,087	1,274	-9	-9
HRU5B	Comm, sand & gravel	1,255	713	1,255	-9	-9
HRU6B	Forest, Till	6,660	6,660	9,371	187,408	1,874
HRU7B	Open, Till	519	519	600	187,408	1,874
HRU8B	Low-resid, Till	7,005	1,875	7,005	-9	-9
HRU9B	High-resid, Till	1,616	1,076	1,616	-9	-9
HRU10B	Forest, Fine deposition	110	110	148	187,408	1,874
HRU11B	Open, Fine deposition	153	153	237	187,408	1,874

In this table, you can enter data for **baseline HRU conditions** and costs associated with land conservation by entering:

- names of the HRUs in your study area,
- acres of each HRU for baseline conditions,
  - Note: "baseline" area can represent the existing conditions or the conditions of future scenario that you would like to model. For example, if you intend to run the model to prioritize management options in 2050, you would enter the projected area of each HRU in 2050.
- "minimum" areas for each HRU – For urban HRUs this may be the existing area of urban HRUs given that these area are not expected to be reforested or otherwise "undeveloped". For forest lands, it may be the area of conserved/protected forest lands which must exist in the future due to their protected status.
- "maximum" areas for each HRU – For urban HRUs, this may be the projected, build-out area or maximum allowable area under zoning regulations. For forest lands, it may be the existing area of forest land given that other HRU types will not be used to regrown forest for urban recreation or start a forestry business.
- cost to conserve HRUs – For example, it may be beneficial to purchase and conserve forest or wetlands. For these HRUs, enter the initial cost of purchasing the land (i.e., capital costs) and any annual operations and maintenance (O&M) costs that may continue to be associated with the purchase.

If land conservation is not possible or desirable for a HRU, then enter "-9" for initial and O&M costs. In the above screenshot example, forest land is possible to conserve at an initial cost of $187,408 per acre and $1,874 annual O&M costs.

Beneath the baseline HRU input table, you will see table(s) for managed HRU sets. Up to 15 sets of **"managed HRUs"** may be specified with corresponding minimum and maximum areas and associated management costs. Enter the name of the management practice in the blue box in the upper right hand corner of the table. The rest of the table is similar to the baseline table. The following input data are requested for each HRU:

- minimum area on which the management practice may be implemented – For urban HRUs, regulations may require that a specific stormwater management practice is implemented.
- maximum area on which the management practice may be implemented – For urban HRUs, some of the HRU may already managed by the specified stormwater practice and is, therefore, unavailable for that treatment.
- initial costs associated with the management practice – For example, design and construction of a bioretention basin to retain one inch of runoff.
- O&M costs associated with the management practice – For example, annual clean out and other upkeep of the bioretention basin to maintain performance.

If a management practice is not applicable or desirable for an HRU, enter "-9" for initial and O&M costs.

First Set of Managed Land Uses and Their Limits				Bioretention basin, 0.6"		< Input name of management practice
HRU ID	Land Use Name	Minimum Area [acre]	Maximum Area [acre]	Initial Cost to Manage [$/acre]	O&M Cost [$/acre/yr]	
HRU1M1	Forest, sand & gravel	0	0	-9	-9	
HRU2M1	Open, sand & gravel	0	0	-9	-9	
HRU3M1	Low-resid, sand & gravel	0	3,099	3,833	38	
HRU4M1	High-resid, sand & gravel	0	1,274	5,685	57	
HRU5M1	Comm, sand & gravel	0	1,255	12,589	126	
HRU6M1	Forest, Till	0	0	-9	-9	
HRU7M1	Open, Till	0	0	-9	-9	
HRU8M1	Low-resid, Till	0	7,005	3,833	38	
HRU9M1	High-resid, Till	0	1,616	5,685	57	
HRU10M1	Forest, Fine deposition	0	0	-9	-9	
HRU11M1	Open, Fine deposition	0	0	-9	-9	

In the above screenshot, all urban HRUs may receive bioretention management. There are no minimum acres of HRU area that must managed but the maximum values are entered based on projected build-out (therefore, same as maximum areas in the baseline table). In addition, as described in the Theoretical Documentation, the maximum area of an HRU that can be managed with bioretention is limited to the area of that HRU that exists considering land conservation decisions (i.e., land area is conserved and no more can be treated than exists as decided is optimal by the model). All specifications are "per acre of HRU"; therefore, the initial cost of $3,833 and O&M cost of $38 for low density residential on sand and gravel surficial geology is the cost to treat one acre of that HRU. The actual footprint of the bioretention basin will only be a small part of that acre of land.

If you have additional managed land use sets, repeat the same instructions for each set. Up to fifteen stormwater management options, including traditional, green infrastructure or LID practices or other land management practices that modify runoff and recharge may be specified. A managed set may include multiple practices that achieve some standard such as retaining a one inch storm event using rooftop disconnection, bioretention basins and swales.

Once this section is complete, navigate to the main screen by pressing "Return to Main":

Return to Main

Check the checkbox next to "Land Use" to indicate that you have completed data entry for this category of input. The button will become gray and help you track which input data are complete.

Next select the "Runoff" button to enter time series data of runoff rates for each HRU:

Selecting "Runoff" brings you to the following input table:

Date (mm/dd/yyyy)	Baseline HRU Set Units: inches/time step										
	HRU1	HRU2	HRU3	HRU4	HRU5	HRU6	HRU7	HRU8	HRU9	HRU10	HRU11
1/1/1989	7.62E-06	1.39E-03	6.54E-05	8.05E-05	2.91E-03	1.78E-03	1.19E-02	7.29E-03	9.47E-03	1.43E-04	4.26E-03
1/2/1989	6.86E-06	1.25E-03	5.76E-05	6.93E-05	2.38E-03	1.60E-03	1.07E-02	6.42E-03	8.15E-03	1.28E-04	3.83E-03
1/3/1989	6.19E-06	1.13E-03	5.07E-05	5.96E-05	1.95E-03	1.44E-03	9.65E-03	5.65E-03	7.01E-03	1.15E-04	3.45E-03
1/4/1989	5.58E-06	1.02E-03	4.47E-05	5.12E-05	1.60E-03	1.29E-03	8.68E-03	4.97E-03	6.03E-03	1.04E-04	3.10E-03
1/5/1989	5.04E-06	9.15E-04	3.93E-05	4.41E-05	1.31E-03	1.16E-03	7.81E-03	4.37E-03	5.18E-03	9.35E-05	2.79E-03
1/6/1989	4.54E-06	8.23E-04	3.46E-05	3.80E-05	1.08E-03	1.05E-03	7.03E-03	3.85E-03	4.46E-03	8.42E-05	2.51E-03
1/7/1989	4.11E-06	7.41E-04	3.05E-05	3.27E-05	8.84E-04	9.44E-04	6.35E-03	3.40E-03	3.86E-03	7.58E-05	2.26E-03
1/8/1989	3.69E-06	6.67E-04	2.68E-05	2.81E-05	7.25E-04	8.49E-04	5.72E-03	3.00E-03	3.33E-03	6.82E-05	2.04E-03
1/9/1989	3.36E-06	6.00E-04	2.36E-05	2.42E-05	5.94E-04	7.65E-04	5.15E-03	2.64E-03	2.86E-03	6.14E-05	1.84E-03
1/10/1989	3.01E-06	5.40E-04	2.08E-05	2.08E-05	4.87E-04	6.88E-04	4.63E-03	2.32E-03	2.46E-03	5.53E-05	1.65E-03
1/11/1989	2.73E-06	4.86E-04	1.83E-05	1.79E-05	4.00E-04	6.19E-04	4.17E-03	2.04E-03	2.12E-03	4.98E-05	1.49E-03

This table requires a time series of runoff rates for baseline and each managed land use set at the daily or monthly time step. For a monthly time step, the day of the month does not matter. The dates entered on sheet will populate the dates in all other input tables that require time series. **Time series data must be consecutive**, that is, there must not be any missing dates. Refer to *Defining Hydrologic Response Units* in *Section 2.1*, for discussion about data sources for runoff and recharge rates.

The time series are input vertically and HRUs and HRU sets horizontally.[22] Therefore to the right of the Baseline HRU set, you will see the continuation of the table shown below.

Managed HRU Set (HRUM1)										
HRU1M1	HRU2M1	HRU3M1	HRU4M1	HRU5M1	HRU6M1	HRU7M1	HRU8M1	HRU9M1	HRU10M1	HRU11M1
7.62E-06	1.39E-03	6.38E-05	6.93E-05	1.07E-03	1.78E-03	1.19E-02	7.11E-03	8.15E-03	1.43E-04	4.26E-03
6.86E-06	1.25E-03	5.61E-05	5.96E-05	8.81E-04	1.60E-03	1.07E-02	6.26E-03	7.01E-03	1.28E-04	3.83E-03
6.19E-06	1.13E-03	4.94E-05	5.12E-05	7.23E-04	1.44E-03	9.65E-03	5.51E-03	6.03E-03	1.15E-04	3.45E-03
5.58E-06	1.02E-03	4.35E-05	4.41E-05	5.93E-04	1.29E-03	8.68E-03	4.85E-03	5.18E-03	1.04E-04	3.10E-03
5.04E-06	9.15E-04	3.84E-05	3.79E-05	4.86E-04	1.16E-03	7.81E-03	4.26E-03	4.46E-03	9.35E-05	2.79E-03
4.54E-06	8.23E-04	3.38E-05	3.26E-05	3.99E-04	1.05E-03	7.03E-03	3.75E-03	3.83E-03	8.42E-05	2.51E-03
4.11E-06	7.41E-04	2.97E-05	2.81E-05	3.27E-04	9.44E-04	6.35E-03	3.31E-03	3.32E-03	7.58E-05	2.26E-03
3.69E-06	6.67E-04	2.61E-05	2.42E-05	2.68E-04	8.49E-04	5.72E-03	2.92E-03	2.86E-03	6.82E-05	2.04E-03
3.36E-06	6.00E-04	2.30E-05	2.08E-05	2.20E-04	7.65E-04	5.15E-03	2.57E-03	2.46E-03	6.14E-05	1.84E-03
3.01E-06	5.40E-04	2.03E-05	1.79E-05	1.80E-04	6.88E-04	4.63E-03	2.26E-03	2.12E-03	5.53E-05	1.65E-03
2.73E-06	4.86E-04	1.79E-05	1.54E-05	1.48E-04	6.19E-04	4.17E-03	1.99E-03	1.82E-03	4.98E-05	1.49E-03

---

[22] If an HRU is excluded from a "managed set" then the values specified are not consequential as the model will exclude using those values.

Once you have entered these data, select "Return to Main" and check the box indicating that this section is complete.

Next select the "Recharge" button to enter time series data of recharge rates for each HRU:

Date	Baseline HRU Set (HRU)										
(mm/dd/yyyy)	HRU1	HRU2	HRU3	HRU4	HRU5	HRU6	HRU7	HRU8	HRU9	HRU10	HRU11
1/1/1989	2.2E-02	5.2E-02	3.4E-02	3.4E-02	3.2E-02	1.2E-02	2.8E-02	2.1E-02	1.9E-02	1.6E-02	3.8E-02
1/2/1989	2.2E-02	5.1E-02	3.4E-02	3.3E-02	3.1E-02	1.2E-02	2.8E-02	2.0E-02	1.8E-02	1.5E-02	3.6E-02
1/3/1989	2.2E-02	5.0E-02	3.3E-02	3.2E-02	3.0E-02	1.2E-02	2.7E-02	2.0E-02	1.8E-02	1.5E-02	3.5E-02
1/4/1989	2.1E-02	4.9E-02	3.2E-02	3.1E-02	2.9E-02	1.2E-02	2.7E-02	1.9E-02	1.8E-02	1.5E-02	3.4E-02
1/5/1989	2.1E-02	4.8E-02	3.1E-02	3.0E-02	2.8E-02	1.2E-02	2.6E-02	1.9E-02	1.7E-02	1.4E-02	3.3E-02
1/6/1989	2.1E-02	4.7E-02	3.1E-02	3.0E-02	2.7E-02	1.1E-02	2.6E-02	1.9E-02	1.7E-02	1.4E-02	3.2E-02
1/7/1989	2.1E-02	4.6E-02	3.0E-02	2.9E-02	2.6E-02	1.1E-02	2.5E-02	1.8E-02	1.7E-02	1.4E-02	3.1E-02
1/8/1989	2.1E-02	4.5E-02	3.0E-02	2.8E-02	2.6E-02	1.1E-02	2.5E-02	1.8E-02	1.6E-02	1.4E-02	3.1E-02
1/9/1989	2.0E-02	4.5E-02	2.9E-02	2.8E-02	2.5E-02	1.1E-02	2.5E-02	1.8E-02	1.6E-02	1.3E-02	3.0E-02
1/10/1989	2.0E-02	4.4E-02	2.9E-02	2.7E-02	2.4E-02	1.1E-02	2.4E-02	1.7E-02	1.6E-02	1.3E-02	2.9E-02
1/11/1989	2.0E-02	4.3E-02	2.8E-02	2.7E-02	2.3E-02	1.1E-02	2.3E-02	1.7E-02	1.5E-02	1.3E-02	2.8E-02

Similar to the runoff input table, the recharge input table also requires a time series of recharge rates for baseline and each managed land use set at the daily or monthly time step. Similarly, it should be consecutive and complete.

Managed HRU Set (HRUM1)										
HRU1M1	HRU2M1	HRU3M1	HRU4M1	HRU5M1	HRU6M1	HRU7M1	HRU8M1	HRU9M1	HRU10M1	HRU11M1
2.2E-02	5.2E-02	3.4E-02	2.9E-02	1.2E-02	1.2E-02	2.8E-02	2.0E-02	1.6E-02	1.6E-02	3.8E-02
2.2E-02	5.1E-02	3.3E-02	2.8E-02	1.2E-02	1.2E-02	2.8E-02	2.0E-02	1.6E-02	1.5E-02	3.6E-02
2.2E-02	5.0E-02	3.2E-02	2.7E-02	1.1E-02	1.2E-02	2.7E-02	1.9E-02	1.6E-02	1.5E-02	3.5E-02
2.1E-02	4.9E-02	3.1E-02	2.7E-02	1.1E-02	1.2E-02	2.7E-02	1.9E-02	1.5E-02	1.5E-02	3.4E-02
2.1E-02	4.8E-02	3.1E-02	2.6E-02	1.0E-02	1.2E-02	2.6E-02	1.8E-02	1.5E-02	1.4E-02	3.3E-02
2.1E-02	4.7E-02	3.0E-02	2.5E-02	1.0E-02	1.1E-02	2.6E-02	1.8E-02	1.5E-02	1.4E-02	3.2E-02
2.1E-02	4.6E-02	3.0E-02	2.8E-02	2.2E-02	1.1E-02	2.5E-02	1.8E-02	1.7E-02	1.4E-02	3.1E-02
2.1E-02	4.5E-02	2.9E-02	2.6E-02	1.7E-02	1.1E-02	2.5E-02	1.8E-02	1.6E-02	1.4E-02	3.1E-02
2.0E-02	4.5E-02	2.8E-02	2.4E-02	9.3E-03	1.1E-02	2.5E-02	1.7E-02	1.4E-02	1.3E-02	3.0E-02
2.0E-02	4.4E-02	2.8E-02	2.3E-02	8.9E-03	1.1E-02	2.4E-02	1.7E-02	1.4E-02	1.3E-02	2.9E-02
2.0E-02	4.3E-02	2.7E-02	2.3E-02	8.6E-03	1.1E-02	2.3E-02	1.7E-02	1.3E-02	1.3E-02	2.8E-02

Select "Return to Main" and check the box indicating that this section is complete.

Step 4. Water Users, Water Demand, Demand Management and Septic System Use. On the Main page, enter the number of water user types. Do not include unaccounted-for-water as it is automatically included in all relevant input tables.

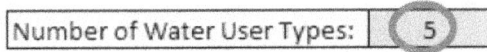

Step 5. Press the "Setup 2" button to automatically prepare input tables for potable, nonpotable, demand management, and septic components of your system. *The process creates blank input tables;*

therefore, do not press this button again unless you have your input data saved elsewhere and want to change the number of water user types.

5. Press "Setup 2" button to prepare input tables for potable and nonpotable demand and septic systems data

Step 6. The "Potable", "Nonpotable" and "Demand Management", and "Septic" buttons will direct you to input tables relating to these input data categories:

Selecting "Potable" will lead you to the following input table:

Date (mm/dd/yyyy)	Total Water Demand [million gallons /time step]					
	Unaccounted	Residential	Commercial	Agricultural	Industrial	Municipal
1/1/1989	0.199	1.937	0.882	0.005	0.008	0.333
1/2/1989	0.199	1.937	0.882	0.005	0.008	0.333
1/3/1989	0.199	1.937	0.882	0.005	0.008	0.333
1/4/1989	0.199	1.937	0.882	0.005	0.008	0.333
1/5/1989	0.199	1.937	0.882	0.005	0.008	0.333
1/6/1989	0.199	1.937	0.882	0.005	0.008	0.333
1/7/1989	0.199	1.937	0.882	0.005	0.008	0.333
1/8/1989	0.199	1.937	0.882	0.005	0.008	0.333
1/9/1989	0.199	1.937	0.882	0.005	0.008	0.333
1/10/1989	0.199	1.937	0.882	0.005	0.008	0.333
1/11/1989	0.199	1.937	0.882	0.005	0.008	0.333

This table requires a time-series of the potable water demand for all users entered in Step 4, plus demand attributable to unaccounted-for-water. This time series should be 1) at the time step of your model, that is, the same time step as runoff and recharge rates, 2) complete and consecutive and 3) the exact same time period as the runoff and recharge rate data.

This section also includes an input table for the average percent consumptive water use by month. These values can reflect any seasonal changes in consumptive use over the year, such as increased outdoor watering in the summer, and among water user types.

Water withdrawal and demand and consumptive use data may be available from state or regional sources. For example, in Massachusetts the Department of Environmental Protection receives such data in the form of Annual Statistical Reports from water utilities.

Month	Average Percent Consumptive Water Use (%)				
	Residential	Commerci	Agricultural	Industrial	Municipal
January	4	4	99	4	4
February	4	4	99	4	4
March	4	4	99	4	4
April	6	6	99	6	6
May	20	20	99	20	20
June	26	26	99	26	26
July	29	29	99	29	29
August	25	25	99	25	25
September	20	20	99	20	20
October	4	4	99	4	4
November	4	4	99	4	4
December	4	4	99	4	4

Note: None of these columns or rows need to add to 100%. Each value is the percent consumptive use for a user type for a month.

Select "Return to Main" and check the box next to the "Potable" button when this section is complete.

Clicking on the "Nonpotable" button will bring you to the following input tables where percent nonpotable water use by user type and percent consumptive nonpotable water use can be filled in with site-specific data. The percent nonpotable water is the maximum amount of potable use that may be met using nonpotable water such as toilet flushing or outdoor irrigation.

The values in the columns or rows do not need to add to 100% for either table.

Month	Maximum Potential Nonpotable Water Use (%)				
	Residential	Commercial	Agricultural	Industrial	Municipal
January	45	90	90	99	90
February	45	90	90	99	90
March	45	90	90	99	90
April	45	90	90	99	90
May	45	90	90	99	90
June	45	90	90	99	90
July	45	90	90	99	90
August	45	90	90	99	90
September	45	90	90	99	90
October	45	90	90	99	90
November	45	90	90	99	90
December	45	90	90	99	90

Month	Average Percent Consumptive Nonpotable Water Use (%)				
	Residential	Commercial	Agricultural	Industrial	Municipal
January	1	1	99	4	1
February	1	1	99	4	1
March	1	1	99	4	1
April	3	3	99	6	3
May	17	17	99	20	17
June	23	23	99	26	23
July	26	26	99	29	26
August	22	22	99	25	22
September	17	17	99	20	17
October	1	1	99	4	1
November	1	1	99	4	1
December	1	1	99	4	1

Based on these nonpotable input data, the consumptive use percent of potable water is recalculated. It is possible to enter values for Maximum Potential Nonpotable Water Use and Average Percent Consumptive Nonpotable Water Use that result in Adjusted Consumptive Potable Water Use values that are outside of the feasible range of 0-100%. To help the user confirm that nonpotable input data do not create infeasible Adjusted Consumptive Potable Water Use values, a third table on the "Nonpotable Demand" worksheet pre-calculates these adjusted values (see below). If any of the values are outside of the feasible range, they are highlighted red. In addition, the model will not run and the user is provided with an error message to change input values for Maximum Percent Nonpotable Use and/or Average Percent Consumptive Nonpotable Water Use. Therefore, ensure that values are not highlighted red in the table shown below before proceeding.

Month	Adjusted Consumptive Potable Water Use (%)				
	Residential	Commercial	Agricultural	Industrial	Municipal
January	6	31	99	4	31
February	6	31	99	4	31
March	6	31	99	4	31
April	8	33	99	6	33
May	22	47	99	20	47
June	28	53	99	26	53
July	31	56	99	29	56
August	27	52	99	25	52
September	22	47	99	20	47
October	6	31	99	4	31
November	6	31	99	4	31
December	6	31	99	4	31

Select "Return to Main" and check the box next to the "Nonpotable" button when this section is complete.

Click on the "Demand Management" button to enter information about how changes in price and other demand management practices may affect demand in your study area.

The first option is reducing demand by increasing the price of water services. Specify the price elasticity – percent change in water use divided by percent change in price – for each type of water user. Price elasticities should be negative given that an increase in price is expected to decrease water use. Price elasticities may be found in the literature but will depend on existing pricing and other local conditions.[23] For example, if the consumer's purchase price of water is relatively high, price elasticities will be smaller than if the existing pricing if relatively low. This reflects the fact that increasing price indefinitely will not decrease demand indefinitely; therefore, it is not a linear effect. The user may specify the maximum price change possible within the planning horizon which may be used to limit price change over the range where the response is expected to be linear.[24]

Price Elasticities [% demand reduction / % price increase]				
Residential	Commercial	Agricultural	Industrial	Municipal
-0.2	-0.2	-0.5	-0.1	-0.2

Initial cost	23,000	$
O&M cost	2,000	$/yr
Maximum price change	20	%

Maximum percent increase in price of water services from existing price over the duration of the planning horizon

The initial cost may reflect the cost of a study to determine effective pricing structure and values, billing frequencies, changes in billing logistics, and consumer outreach to convey the importance of efficient use of water resources and the planned change in pricing. O&M costs may reflect smaller studies to re-evaluate pricing every year or five years; however, be sure to enter the expected *annual* cost of such evaluations.

The second option is direct demand reductions which may be achieved using rebates for water efficient appliances, changing building codes, educational outreach and other practices. Initial and O&M costs may be specified for the aggregate cost of direct demand reduction practices. The aggregate effect of these practices should be specified as a percent reduction is overall demand.

Initial cost	3,186,600	$
O&M cost	0	$/yr
Total demand reduction	0.60	MGD

Total demand reduction value should equal the MGD reduction in demand across all user types achieved by all managemnet practices encompassed in the initial and O&M cost.

EPA's WaterSense website provides a calculator that together with local or Census data (e.g., number of households) can be used to determine the total potential reductions in water use with the

---

[23] For example, http://www.hks.harvard.edu/fs/rstavins/Monographs_&_Reports/Pioneer_Olmstead_Stavins_Water.pdf

[24] The effect of price on water is assumed to be linear with WMOST v1 but nonlinear assumption may be implemented in future version.

installation of water efficient appliances.[25] **When acquiring input data for these practices, the user must be aware of the potential reduction in the individual effectiveness of demand management practices when multiple practices are implemented simultaneously.[26]**

For any options that are not possible or desirable, enter -9 for costs.

Select "Return to Main" and check the box next to "Demand Management" when this section is complete.

Click on the "Septic" button to enter information about the percent of customers with septic systems inside and outside of your study area *that are on public water*. Customers that are not on public water should be represented as private withdrawals and discharges on the Surface Water or Groundwater input worksheets depending on their source and discharge of water (see Step 7 below for description of these input worksheets).

For public water users, it is important to distinguish customers who are on septic systems but are outside of the watershed of the study area being modeled. Such septic systems do not recharge the groundwater and do not contribute to the baseflow of the stream in the study area's watershed.

Customers with Public Water & Septic Systems Recharging Inside Study Area (%)				
Residential	Commercial	Agricultural	Industrial	Municipal
9.4	9.4	9.4	9.4	9.4

Customers with Public Water & Septic Systems Recharging Outside Study Area (%)				
Residential	Commercial	Agricultural	Industrial	Municipal
0	0	0	0	0

Select "Return to Main" and check the box next to "Septic" when this section is complete.

Step 7. Surface Water, Groundwater, Interbasin Transfer and Infrastructure. Here you will see buttons which will bring you to the Surface Water, Groundwater, Interbasin Transfer, and Infrastructure input tables:

☐ Surface Water & In-Stream Flow Targets   ☐ Groundwater   ☐ Interbasin Transfer   ☐ Infrastructure

Clicking on the "Surface Water" button will bring you to three input tables.

In Part 1 of this section, you can enter reservoir or surface storage properties and costs. Reservoir and surface storage may represent reservoirs, lakes or ponds used for water supply and/or surface storage tanks. Surface storage in wetlands may be modeled as surface storage or as a separate HRU. Initial

---

[25] http://www.epa.gov/watersense/our_water/start_saving.html#tabs-3

[26] For example, rebates for water low flow shower heads will reduce the gallons per minute used in showering. If an increase in water rates is implemented at the same time, the anticipated water use reduction may not be as large with a low flow shower head as with a high flow shower head even if the new water rates induce shorter shower times.

volume is the volume at the start of modeling period. Minimum target volume may represent the volume of water always maintained in storage for emergencies or inactive storage volume which is inaccessible due to the height of the storage outlet. Existing maximum volume is the total volume of existing storage. Initial costs should include the cost to plan, design and build additional surface storage volume. O&M costs should include the annual cost for maintaining surface storage capacity in operational condition.

Initial reservoir/surface storage volume	533	[MG]
Minimum target reservoir/storage volume	0.0	[MG]
Existing maximum reservoir/storage volume	710	[MG]
Initial construction cost	1,542,790	[$/MG]
O&M costs	0	[$/MG]

To exclude an increase reservoir/surface storage volume as a management option, enter -9 in the input field shown below.

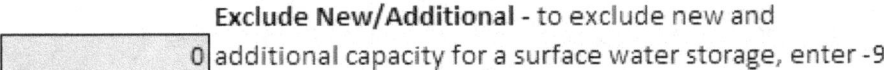

Exclude New/Additional - to exclude new and additional capacity for a surface water storage, enter -9

0

In Part 2 you may enter information about private withdrawals and discharges of surface water such as industrial users that are not on public water. These data may be available from state sources such as the Department of Environmental Protection or regional sources such as regional EPA offices. In addition, if the stream into which your study area drains receives inflow from an upstream reach, enter a time series for the inflow of this surface water. These data should be available from the model from which you may have obtained your RRRs. These time-series must be at the resolution of your model (i.e., daily or monthly) and over the same time period as other time series. The dates will be pre-filled for you based on data you entered in the Runoff tab. As with other time series data, they must be complete and consecutive. For any of the three time series, if you do not have data or they do not exist, enter zero for all dates. Note that upstream inflow is critical, especially if you will be specifying any streamflow requirements.

For withdrawals and discharges that do not exist, enter 0.

Date (mm/dd/yyyy)	Private Sw Withdrawal [MG/time step]	Private Sw Discharge [MG/time step]	External Sw Inflow [cfs]
1/1/1989	0.00	0.00	19.91
1/2/1989	0.00	0.00	18.60
1/3/1989	0.00	0.00	17.80
1/4/1989	0.00	0.00	17.09
1/5/1989	0.00	0.00	16.23
1/6/1989	0.00	0.00	15.53
1/7/1989	0.00	0.00	14.92
1/8/1989	0.00	0.00	14.42
1/9/1989	0.00	0.00	13.92
1/10/1989	0.00	0.00	13.31
1/11/1989	0.00	0.00	12.60

In Part 3 you may provide management goals for minimum and/or maximum in-stream flow on a monthly basis. In addition, any requirements for flow to a downstream reach may be specified. Requirements or guidelines for minimum and/or maximum in-stream flow may be found at the state or regional level. For example, in New England there are Stream Flow Recommendations[27] and in Massachusetts there is a Sustainable Water Management Initiative Framework.[28] If any of these flow requirements do not exist in your study area, enter "-9" for each month of that set.

---

[27] http://www.fws.gov/newengland/pdfs/Flowpolicy.pdf

[28] http://www.mass.gov/eea/docs/eea/water/swmi-framework-nov-2012.pdf

For minimum and maximum values, enter -9 and the model will not apply the constraint

Month	Minimum In-Stream Flow [cfs]	Maximum In-stream flow [cfs]	Minimum Sw Outflow to External Sw [cfs]
January	16.56	-9	-9
February	19.10	-9	-9
March	17.27	-9	-9
April	19.46	-9	-9
May	15.98	-9	-9
June	18.50	-9	-9
July	18.46	-9	-9
August	18.75	-9	-9
September	18.85	-9	-9
October	17.51	-9	-9
November	17.51	-9	-9
December	17.09	-9	-9

Select "Return to Main" and check the box next to "Surface Water" when this section is complete.

Clicking on the "Groundwater" button will direct you to three input tables.

As in the Surface Water section, the Groundwater input tables consist of three parts. The same state and regional data sources are recommended as for surface water data. In Part 1, enter information about groundwater storage characteristics will likely be derived from the same model that you obtain the runoff and recharge rates. These data include:

- groundwater recession coefficient or baseflow coefficient – fraction of groundwater volume that flows to the stream reach each time step,
- initial groundwater volume – volume of the active groundwater aquifer at the start of the modeling period,
- minimum volume – this volume may be based on the depth of wells which are used for water supply below which water is inaccessible and/or the volume at which the water table will be below the stream bed and therefore no longer emptying to the stream, and
- maximum volume – this value represents the total storage capacity of the aquifer.

Groundwater recession coefficient	0.01	[1/time step]
Initial groundwater volume	1,134	[MG]
Minimum volume	706	[MG]
Maximum volume	2,838	[MG]

In Part 2, similar to the Surface Water tab, you can enter time series data for private groundwater withdrawals, discharges and inflow into the study area.

For withdrawals and discharges that do not exist, enter 0.

Date [mm/dd/yyyy]	Private Gw Withdrawal [MG/time step]	Private Gw Discharge [MG/time step]	External Gw Inflow [MG/time step]
1/1/1989	0.00	0.00	0.00
1/2/1989	0.00	0.00	0.00
1/3/1989	0.00	0.00	0.00
1/4/1989	0.00	0.00	0.00
1/5/1989	0.00	0.00	0.00
1/6/1989	0.00	0.00	0.00
1/7/1989	0.00	0.00	0.00
1/8/1989	0.00	0.00	0.00
1/9/1989	0.00	0.00	0.00
1/10/1989	0.00	0.00	0.00
1/11/1989	0.00	0.00	0.00

In Part 3, similar to the Surface Water tab, you can enter requirements for groundwater flowing out of the basin. In most cases this will not exist as the groundwater will drain to the stream reach; however, this option provides flexibility in defining a study area or when groundwater and surface water watersheds do not overlap.

For minimum and maximum values, enter -9 and the model will not apply the constraint

Month	Minimum External Gw Outflow [MG/time step]
January	-9.00
February	-9.00
March	-9.00
April	-9.00
May	-9.00
June	-9.00
July	-9.00
August	-9.00
September	-9.00
October	-9.00
November	-9.00
December	-9.00

Select Return to Main and check the box next to "Groundwater" when this section is complete.

Clicking the "Interbasin Transfer" (IBT) button will lead you to the two sets of input data.

In Part 1, you can enter data for:

- costs to purchase water and wastewater from systems outside of your study area and
- initial costs for water and wastewater rights in addition to any existing agreements including costs for any new infrastructure to utilize the additional rights.[29]

If you do not want IBT as a management option, enter -9 for costs AND 0 for constraints.

Purchase cost for potable water	3,803	[$/MG]
Purchase cost for wastewater	6,340	[$/MG]

Initial cost for new/increased IBT potable water limit	29,500,000	[$/MGD]
Initial cost for new/increased IBT wastewater limit	0	[$/MGD]

In Part 2, enter any existing monthly limits for interbasin transfer of water and wastewater in the left and daily or annual limits in the right table. Depending on the time step of your model, the daily, monthly and/or annual limits are adjusted to specify appropriate constraints in the model.

Enter existing limits on IBT for daily, monthly and/or annual basis. If a constraint does not exist, enter -9.

Month	Existing Limits on IBT [MG per month]	
	Water	Wastewater
January	-9.00	-9.00
February	-9.00	-9.00
March	-9.00	-9.00
April	-9.00	-9.00
May	-9.00	-9.00
June	-9.00	-9.00
July	-9.00	-9.00
August	-9.00	-9.00
September	-9.00	-9.00
October	-9.00	-9.00
November	-9.00	-9.00
December	-9.00	-9.00

Existing Limits on IBT	Water	Wastewater
Daily [MGD]	-9.00	6.00
Annual [MG per year]	-9.00	-9.00

Additional Capacity Limits	Water	Wastewater
Daily [MGD]	0.27	-9.00

The following guidelines for specifying limits and initial costs for increasing limits are important to note:

- **If you do not have interbasin transfer as an option, you must enter "0" for limits.** Entering "-9" will indicate no restriction, that is, unlimited interbasin transfer is available. As such if you enter -9 for daily, monthly or annual limits, then you must specify the initial cost for new/increased IBT.

- If additional water or wastewater services can be purchased with **no additional initial costs or entry fees, then enter the current agreement limit for services and specify $0 for initial cost** for a new/increased limit (i.e., do not enter -9 for the existing limit).

---

[29] The second case study of Danvers and Middleton, MA provides costs associated with initial connection for water with the Massachusetts Water Resources Authority, a large regional water and wastewater provider.

- If your system provides water services to customers outside of the basin without a return flow via the wastewater treatment plant or septic systems, you may specify these customers as a separate water user type that entirely drains to septic outside of the study area. If your system provides out of basin wastewater services that discharge in your basin, you may enter this flow as a private discharge of surface water (or groundwater, depending on where the wastewater treatment plant discharges). WMOST v1 does not support routing out of basin wastewater to the wastewater treatment plant. It may be added as functionality in future versions.

- If your system's wastewater is treated outside of the basin at a larger, central facility and you want to model returning the treated wastewater for discharge locally, then you may enter a capital cost for a wastewater treatment plant that represents the construction of infrastructure necessary to return and discharge the treated wastewater. In addition, enter O&M costs that reflect the IBT O&M cost and exclude the use of IBT for wastewater. This will effectively model the desired scenario. If the returned wastewater will be discharged to groundwater rather than surface water, follow the same procedure but apply it to the aquifer storage and recharge facility rather than the wastewater treatment plant. See below under "Infrastructure" for input data tables related to wastewater treatment plant and aquifer and storage recharge facility.

Select "Return to Main" and check the box next to "Interbasin Transfer" when this section is complete.

Clicking "Infrastructure" will lead you to the next section, where you can add information about costs and capacity limits for a range of water and wastewater facilities. This section consists of six parts.

In Part 1, you enter the planning horizon for large capital improvement projects and the interest rate for loans for such projects. For any management option for which a project lifetime is not requested, the planning horizon is used for the lifetime over which the initial cost is annualized. The specified interest rate is used for the annualization of all initial and capital costs. For mathematical equations describing the annualization of capital costs, please refer to Section 2.1.1 in the separate Theoretical Documentation.

Planning horizon [years]	20
Interest rate [%]	5.00

In Part 2, you enter data related to providing water services including:

- Consumer's price for potable water – this may be specified as a monthly fixed fee and/or volume based fee,
- Facility data for groundwater pumping, surface water pumping and water treatment plant including
  - Capital costs – cost for increasing capacity or cost for replacing existing capacity beyond the remaining lifetime,
  - O&M costs – cost for operating based on the size and flow through the facility,
  - Existing maximum capacity of the facility,
  - Lifetime remaining on existing infrastructure or the number of years expected to remain before major capital rehabilitation or new facility must be built, and

- o Lifetime of new infrastructure – the expected lifetime of new construction before major capital rehabilitation or new facility must be built,
- Potable distribution system data including
  - o Initial cost for surveying the distribution system for leaks and repairing to the maximum percent feasible,
  - o O&M costs representing annual costs for maintaining repairs made to the distribution system, and
  - o Maximum percent of distribution system leaks that can be fixed – this value may be less than 100% due to practical limitations of many miles of pipes.

If no water treatment plant exists in your study area (i.e., all water is from interbasin transfer), then enter "0" for maximum capacities and remaining lifetimes. However, still enter the price that is charged for customers for water services. To exclude the option to increase facility capacity, enter -9 in the "Exclude New/Additional" for the appropriate facility.

Water services	Value	Units	Exclude New/Additional
Consumer's price for potable water: Fixed fee	0	$/month	
Consumer's price for potable water: Variable, volume-based fee	5.03	$/HCF	
Groundwater (Gw) Pumping			0
Capital cost for additional capacity	747,285	$/MGD	
Operation & Maintanance (O&M) costs	-	$/MG	
Existing maximum capacity	1.74	MGD	
Lifetime remaining on existing infrastructure	33	yrs	
Lifetime of new construction	35	yrs	
Surface Water (Sw) Pumping			0
Capital cost for additional capacity	453,885	$/MGD	
O&M costs	-	$/MG	
Existing maximum capacity	9.40	MGD	
Lifetime remaining on existing infrastructure	33	yrs	
Lifetime of new construction	35	yrs	
Water Treatment Plant (WTP)			0
Capital cost for additional capacity	2,022,884	$/MGD	
O&M costs	5,314	$/MG	
Existing maximum capacity	9.40	MGD	
Lifetime remaining on existing infrastructure	33	yrs	
Lifetime of new construction	35	yrs	
Unaccounted-for-Water/ Potable water distribution system leak			
Initial cost for survey & repair	774,368	$	
O&M costs for maintaining reduction in UAW	77,437	$/yr	
Maximum percent UAW that can be fixed	99	%	

In Part 3, enter similar data for wastewater services as for water services including consumer's price, capital and O&M costs, lifetime of new and existing infrastructure, and repair of infiltration into collection system. Two additional data are requested:

- "Are wastewater fees charged based on metered water or wastewater?" –Most wastewater utilities in the U.S. charge for wastewater services based on metered potable water delivered to a customer. However, the option is provided to charge based on metered wastewater determine the effect of separating metering.
- "Existing Gw infiltration into collection system" – Specify the percent of wastewater inflow to the wastewater treatment plant that is groundwater infiltration.

Wastewater treatment plant (WWTP)	Value	Units	Exclude New/Additional
Consumer's price for wastewater services: Fixed fee	0.00	$/month	0
Consumer's price for wastewater services: Variable, volume-based fee	6.12	$/HCF	
Are wastewater fees charged based on metered water or wastewater?	water	water or wastewater	
Capital cost for additional capacity	15,788,674	$/MGD	
O&M costs	7,925	$/MG	
Existing maximum capacity	0.00	MGD	
Lifetime remaining on existing infrastructure	0	years	
Lifetime of new construction	40	years	
Infiltration into wastewater collection system			
Existing Gw infiltration into collection system	0	% of WW Inflow	
Initial cost for survey & repair	214,846	$	
O&M costs for maintaining reduction in infiltration	21,485	$/yr	
Maximum percent of infiltration that can be fixed	0	%	

To exclude the option to increase wastewater treatment plant capacity, enter -9 in the "Exclude New/Additional" data field.

In Part 4, enter data for a water reuse facility (WRF) similar to water and wastewater facilities including the ability to exclude new and additional capacity.

Water Reuse Facility (WRF)	Value	Units	Exclude New/Additional
Capital cost for additional/ new capacity	10,402,467	$/MGD	0
O&M costs	2,850	$/MG	
Existing maximum capacity	0.00	MGD	
Lifetime remaining on existing infrastructure	0.00	yrs	
Lifetime of new construction	35	yrs	

In Part 5, enter data for a nonpotable water distribution system which are similar to the other facilities but in addition, specify the price that would be charged to customers for the provision of nonpotable water. See case study appendices for potential data sources.

Nonpotable Water Distribution System	Value	Units	Exclude New/Additional
Consumer's price for nonpotable water: Fixed fee	0	$/month	0
Consumer's price for nonpotable water: Variable, volume-based fee	3.02	$/HCF	
Capital cost for additional capacity	12,529,440	$/MGD	
O&M costs	1,716	$/MG	
Existing maximum capacity	0.00	MGD	
Lifetime remaining on existing infrastructure	0	yrs	
Lifetime of new construction	35	yrs	

In Part 6, enter data for an aquifer storage and recovery (ASR) facility similar to the other facilities.

Aquifer Storage and Recovery (ASR)	Value	Units	Exclude New/Additional
Capital cost for additional/new capacity	10,807,824	$/MGD	0
O&M costs	3,769	$/MG	
Existing maximum capacity	0	MGD	
Lifetime remaining on existing infrastructure	0	yrs	
Lifetime of new construction	35	yrs	

Select "Return to Main" and check the box next to "Infrastructure" when this section is complete.

Step 8. **Measured In-stream Flow.** Click on the "Measured Flow" button to lead you to the next input table. These data are used to create an output graph showing both measured and modeled in-stream flow to assess the accuracy of the model in reproducing measured flows. These data may be acquired from the U.S. Geological Survey or from the model from which you may have obtained RRRs.

Date (mm/dd/yyyy)	Measured In-Stream Flow (cfs)
1/1/1989	25.66
1/2/1989	26.77
1/3/1989	26.27
1/4/1989	25.06
1/5/1989	23.90
1/6/1989	22.79
1/7/1989	21.63
1/8/1989	20.72
1/9/1989	19.91
1/10/1989	19.06
1/11/1989	18.10

Select "Return to Main" and check the box next to "Measured Flow" when this section is complete.

Once all sections are complete, you may run the optimization model by clicking the red "Optimize" button. This will initiate the optimization and processing of results.

RUN OPTIMIZATION

Optimize

Once the optimization is complete, the model will display the message below. Click "OK" and wait for the model to process outputs and populate the Results tables. The Main page will display again once the output processing is complete.

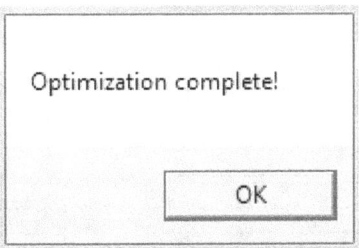

Optimization complete!

OK

2.3   Evaluating Results

After optimization, WMOST provides three outputs:

1. summary table of management practices and associated costs that met specified goals (e.g., minimum demand, minimum in-stream flow) at least cost,
2. graph of modeled in-stream flow and baseflow compared with user-specified measured in-stream flow, and
3. modeled in-stream flow and baseflow and user-specified minimum and/or maximum in-stream flow targets.

Results represent estimated conditions at end of the planning horizon if all management practices were implemented. For example, the modeled in-stream flow and baseflow are estimated to occur if recommended management practices are implemented and human demand is at the projected rate input by the user with the expected weather patterns (i.e., user input runoff and recharge rates). The flows over the modeling period represent estimated flow over a variety of potential weather conditions represented by the years in the modeling period. The length of the modeling period and the variety of conditions it represents determines the robustness and sustainability of the solutions recommended by the model. In addition, **performing sensitivity analyses is highly recommended** especially for input data with least certainty to further determine the robustness of results. By varying the input data, you can determine the robustness of results over a variety of potential conditions that may occur by the end of planning period.

To view the summery table of results, select the "Results Table" button to display the management decisions and associated costs. Capital and O&M costs are presented as one total annualized cost in WMOST v1. This may lead to costs for an existing facility even if no additional capacity is selected as a management practice. For example, an existing water treatment plant may be able to meet projected demand without additional capacity but O&M costs are still incurred for operating the facility for the required demand. Therefore, when "number of units" is zero but there is still a cost, that cost represents O&M costs.

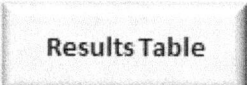

An excerpt from the summary table of results is below:

RESULTS

Due to solver precision, there may be neglible changes in HRU areas that round to zero displayed as 0 or (0).

Total Annual Cost	$13.5	million
Water Revenue	$10.2	million
Wastewater Revenue	$10.3	million

MANAGEMENT PRACTICES	UNITS	Number of Units	Total Annual Sub-Costs (incl. O&M)
Demand Management			
Consumer Rate Change	%	20	$3,846
Direct Demand Reduction	MGD	0.60	$255,701
Land Conservation			$0
Forest, sand & gravel	Acres	0	
Open, sand & gravel	Acres	0	
Low-resid, sand & gravel	Acres	(0)	
High-resid, sand & gravel	Acres	0	

Select the "Compare to Measured Flow" button to display a graph comparing measured in-stream flow to modeled in-stream flow and baseline.

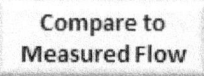

Select the "Compare to Target Flow" button to display a graph comparing modeled flows to user-specified in-stream flow constraints.

Compare to
Target Flow

Results may be printed from the Excel interface with the same options as any Excel file or copied and pasted into Word or another application.

2.4 User Tips

The following tips are provided for troubleshooting, interpreting results and modeling specific situations or scenarios.

- If the results show "1E+30" for Total Annual Cost, **the scenario run was infeasible.** This means that the specified management goals and/or continuity constraints could not be met with the user-provided input data. Refer to the Theoretical Documentation for constraints that are defined in the optimization model. You may need to adjust your management goals or identify erroneous input data. Future versions of the model will support identifying constraints and data that contribute to infeasible solutions.

- If you want to test the effect of a management option but the model is not selecting it, you can enter 0 for cost. You can also adjust the cost of a management practice to see the cost at which that practice is selecting by the model and, therefore, assessed as cost effective.

- To exclude replacement costs for existing infrastructure, set the remaining lifetime to be greater than the planning period. This tells the model that the infrastructure does not need replacement within the planning period and the model will not calculate replacement costs. It will only calculate capital costs for new or additional capacity of infrastructure and O&M costs.

- As detailed in the Theoretical Documentation, a "sustainability" constraint forces the initial and final groundwater and reservoir/surface water storage volumes to be equal. If only one year is modeled, then the watershed should be a "within-year" watershed, that is, the groundwater and reservoir volumes generally return to their initial levels each year. If multiple years are modeled, this sustainability constraint "softens" and the model may be applied to multi-year watersheds as well.

- A "simulation run" is advised before optimization runs to determine the accuracy of WMOST in modeling in-stream flow relative to measured data or data from the detailed watershed simulation model from which RRRs may have been acquired. *Section* 3.2.2 describes the process for performing a "simulation run" with WMOST.

- Sensitivity analyses should be performed with the most uncertain input data. For example, if the price elasticity for industrial water use is most uncertain, then the model should be run multiple times over a range of potential values as follows:

  1. Starting with the best estimated value, determine the range of potential values e.g., -0.5 with a potential range of -0.2 to -0.7.

  2. Run the model with the same input data varying only the price elasticity for industrial water use. For example, run the model five times with values of -0.2, -0.3, -0.5, -0.6, and -0.7.

3. Save the results of each run, that is, either use the "save as" function in Excel to save a different version of the file/model with each run or copy and paste the results tables into a separate Excel file.

4. Determine the effect of the price elasticity on results. Does it change whether demand management via pricing is implemented? Does it change the mix of other management options? How does it change the total annual cost?

Ideally, change only one of the input data at a time at first so that you can determine the individual effect of each variable. Once you know the individual effects and you have more than one uncertain input data, you may want to run the model varying more than one data at a time to determine their combined effects. You may consider "worst" and "best" case scenarios. For example, vary all uncertain data in the direction of higher costs to determine the worst case scenario for total cost if all uncertain data were to be truly in the higher cost direction. Or run the highest cost for a specific management practice to determine the whether it is still a cost effective practice that is chosen by the model and, therefore, a "no regrets" option. For more guidance, please refer to EPA's "Sensitivity and Uncertainty Analyses" website.[30]

- Trade-off analyses are similar to sensitivity analyses but with a different purpose. With trade-off analyses the question may be "How does cost change with increasing in-stream flow? Is it linear? Are there points at which the increasing investment in management practices (i.e., total cost) results in less increase in in-stream flow than the first $X?" To answer these questions, follow the same steps as for the sensitivity analysis. For the in-stream flow example, increase the minimum in-stream flow requirement with each run and record the results. Then examine the effect of this increase on the combination of management practices that are suggested and the total costs and revenues. A trade-off curve may be created, as in *Section 3.2.3*, by plotting total cost versus percent of in-stream flow requirement to create a visual understanding of the trade-off and results.

---

[30] http://www.epa.gov/osa/crem/training/module8.htm

3. Case Study Examples

3.1 Upper Ipswich River Basin

The Ipswich River Basin (IRB) in Massachusetts is used as a case study for the application of the model. The upper IRB is the watershed of the South Middleton Gaging Station of the United States Geological Survey (USGS) on the Ipswich River and experiences low and no flow events during summer months (*Exhibit 3*). The model is applied to the upper IRB to evaluate a broad range of management options for meeting these objectives. A detailed modeling study of the IRB watershed system was conducted by Zarriello and Ries (2000)[31] of the USGS. That study compiled extensive information and data on the basin which were used here. Relevant background information is summarized below and the reader is referred to the 2000 study for a detailed watershed description.

Exhibit 3. Map of the Upper Ipswich River Basin.

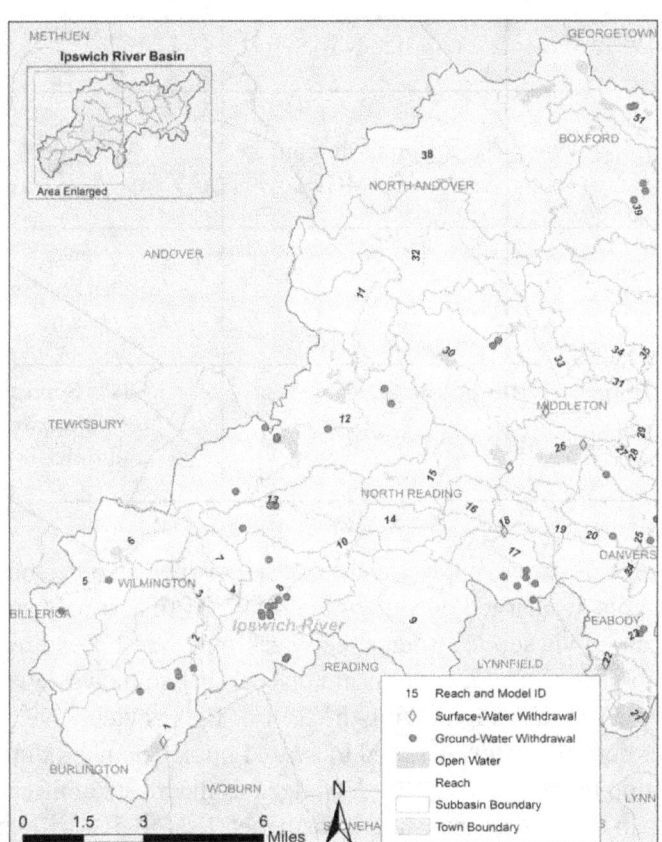

The upper IRB covers approximately 44 square miles (~Hydrologic Unit Code (HUC)-12 sub-watershed) out of the total IRB area of approximately 150 square miles (~HUC-10 watershed). Of

---

[31] Zarriello, P. J. and Ries III, K. G. (2000). A Precipitation-Runoff Model for Analysis of the Effects of Water Withdrawals on Streamflow, Ipswich River Basin, Massachusetts. United States Geological Survey Water-Resources Investigation Report 00-4029.

this land area, 77% is developed. It comprises 14 towns but only four of these towns, Reading, North Reading, Wilmington and Lynnfield, utilize the upper IRB for their water supply. The town of Lynn is not located in the upper IRB but obtains 16% of its water supply from it (Zarriello and Ries 2000). *Exhibit 4* below lists the percent of each town's area within the upper IRB, percent of water supply obtained from the IRB and resources for water and wastewater withdrawal and discharge.

Exhibit 4. Upper IRB Water Supply and Wastewater Services

Town	Area in upper IRB	Supply from upper IRB	Water Source	Wastewater Discharge
Lynn	0%	16%	Sw	Sewer (discharges out of basin)
Lynnfield	32%	100%	Gw (April to Nov-Sw)	Septic
North Reading	100%	100%	Gw (summer-import <1.5 MGD)	Septic
Reading	48%	100%	Gw	Sewer (discharges out of basin)
Wilmington	83%	100%	Gw	84% Septic (16% discharges out of basin)

Groundwater is almost exclusively the source of water supply except for Lynn which also lies outside of the basin. The majority of the wastewater is discharged outside of the basin. At first, it may appear that the majority of the wastewater is recharged via septic systems; however, only North Reading is entirely within the basin boundary. Therefore, even septic systems are discharging some wastewater to other basins and are neither recharging the IRB nor augmenting the flow of the Ipswich River. Extensive groundwater withdrawals, the export of wastewater, and increased human demand during low precipitation and high evapotranspiration months have been recognized as the most significant contributors to the low and no flow events in the late summer in the basin (see *Exhibit 5*, Zarriello and Ries 2000). *Exhibit 5* shows in bold the three most critical months in 1999 when the lowest percent of target flows were met and human demand were also the highest all year.

Exhibit 5. Hydrologic Conditions in 1999

Date	Target Flow (ft^3/s)	Streamflow (ft^3/s)	Streamflow (as % of Target)	Precipitation (in/month)	Human Demand (ft^3/s)
Janurary	44	106.2	240%	6.9	8.4
February	44	135.0	305%	4.5	8.2
March	111	128.9	116%	4.0	8.1
April	111	49.1	44%	0.9	8.9
May	66	30.6	46%	3.3	10.3
June	**22**	**4.7**	**22%**	**0.1**	**11.1**
July	**22**	**0.9**	**4%**	**4.7**	**11.6**
August	**22**	**0.2**	**1%**	**1.5**	**10.9**
September	22	19.1	88%	9.3	10.3
October	22	32.7	151%	4.9	9.0
November	44	46.1	104%	2.4	8.3
December	44	48.3	109%	2.3	9.8

Note: Bold rows highlight the most critical months in 1999 when the lowest percent of target flow were met and human demand were highest (Data from Zarriello, 2002[32] and Zarriello and Ries, 2000)

Optimization Scenario

WMOST was configured for a monthly time step for one year of modeling. We compiled data from the USGS model as well as local sources as documented in Zoltay (2007). The data were for the year 1999 because it was the latest of the years for which pumping data were based on measured data in the USGS model. The following management practices were specified as available for meeting U.S. Fish and Wildlife in-stream flow targets and 1999 human water demand:

- Increasing or building new capacity for surface water pumping, groundwater pumping, water treatment plant, wastewater treatment plant, water reuse facility, aquifer storage and recharge facility, and nonpotable distribution system,

- Repairing leaks in the potable water distribution system and infiltration and inflow in the wastewater collection system,

- Demand management via changes in pricing of water services,

- Bioretention basin for all HRUs except forest, and

- Land conservation of forest HRU.

Interbasin transfers were excluded for the example scenario documented in this User Guide.

Results

Exhibit 6 summarizes the management options recommended by the model, along with the sub-costs and total cost. The solution includes wastewater treatment capacity because wastewater services were

---

[32] Zarriello, P. J. (2002). Simulation of Reservoir Storage and Firm Yields of Three Surface-Water Supplies, Ipswich River Basin, Massachusetts. United States Geological Survey Water-Resources Investigation Report 02-4278.

previously outsourced. In addition, the model "used" additional surface water pumping and aquifer storage and recharge to shift the timing of surface and groundwater withdrawals and the recharge to the aquifer which, after a delay, discharges as baseflow to the stream. In addition, reduction in demand was implemented by increasing the consumer's price for water and eliminating leakage from the distribution system. Finally, wastewater treatment costs and the loss of groundwater were minimized by reducing groundwater infiltration into the sewer collection system.

Exhibit 6. Results for Meeting Minimum In-Stream Flow and Human Demand

Total Annual Cost	$31.7	million
Water Revenue	$13.1	million
Wastewater Revenue	$3.4	million

MANAGEMENT PRACTICES	UNITS	Number of Units	Total Annual Sub-Costs (incl. O&M)
Consumer Rate Change	%	50	$3,260
Additional SW Pumping Capacity	MGD	165	$1,925,580
Additional GW Pumping Capacity	MGD	0	$210,649
Additional Surface Water Storage	MGD	0	$1,064,000
Additional WTP Capacity	MGD	0	$13,850,100
Potable Distribution System Repair	% of Leaks	100	$70,423
Additional WWTP Capacity	MGD	57	$9,392,270
Infiltration Repair	% of Infiltration	100	$154,777
Additional ASR Capacity	MGD	281	$4,997,340

Compared to measured flows for a monthly model, to in-stream flows appear reasonable for both magnitude and behavior or pattern.

Measured and Modeled Flows

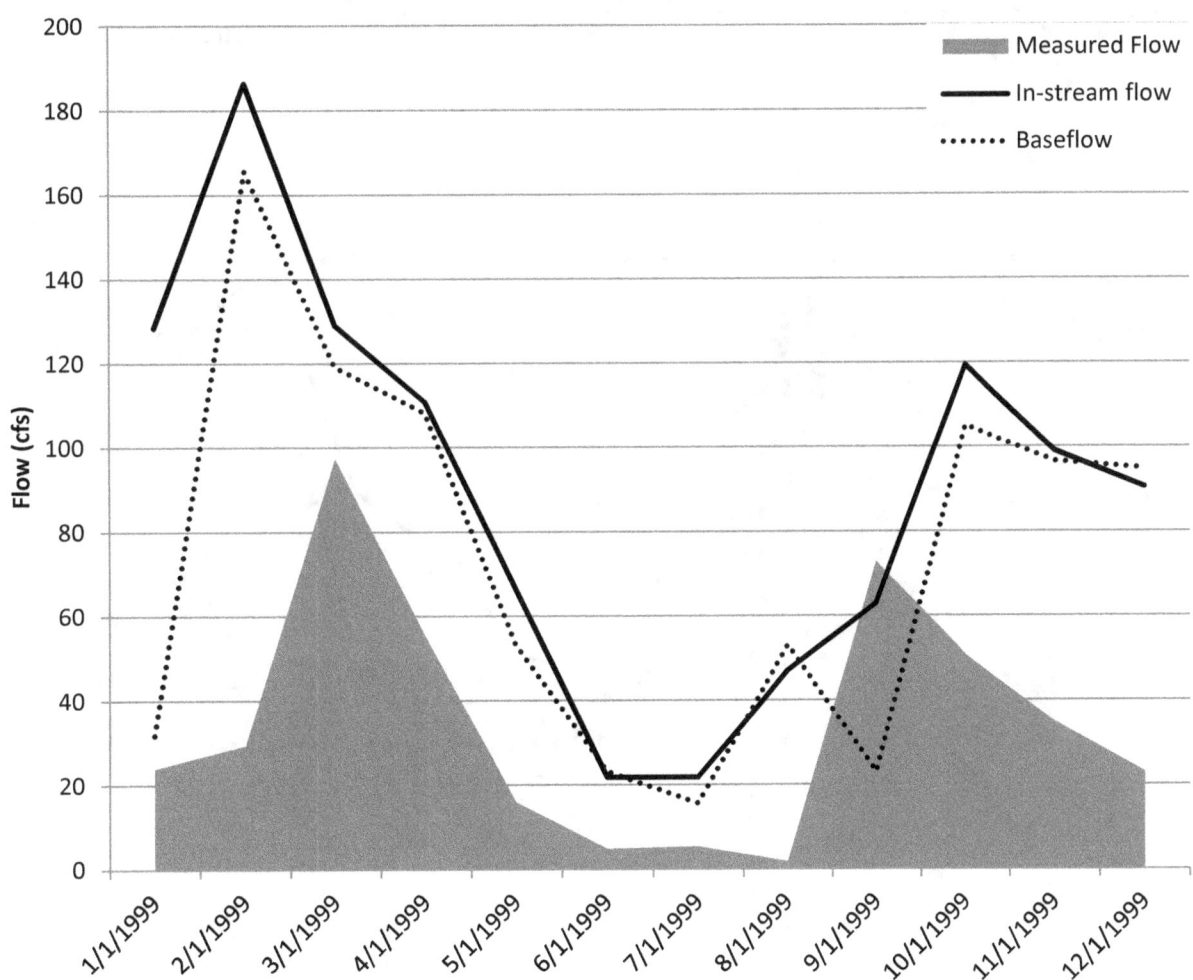

Comparing the modeled in-stream flow to the specified target flows (i.e., minimum in-stream flow), it is clear where the target 'forced' the model to implement management options to increase in-stream flow (i.e., where the modeled in-stream flow is adjacent to the target).

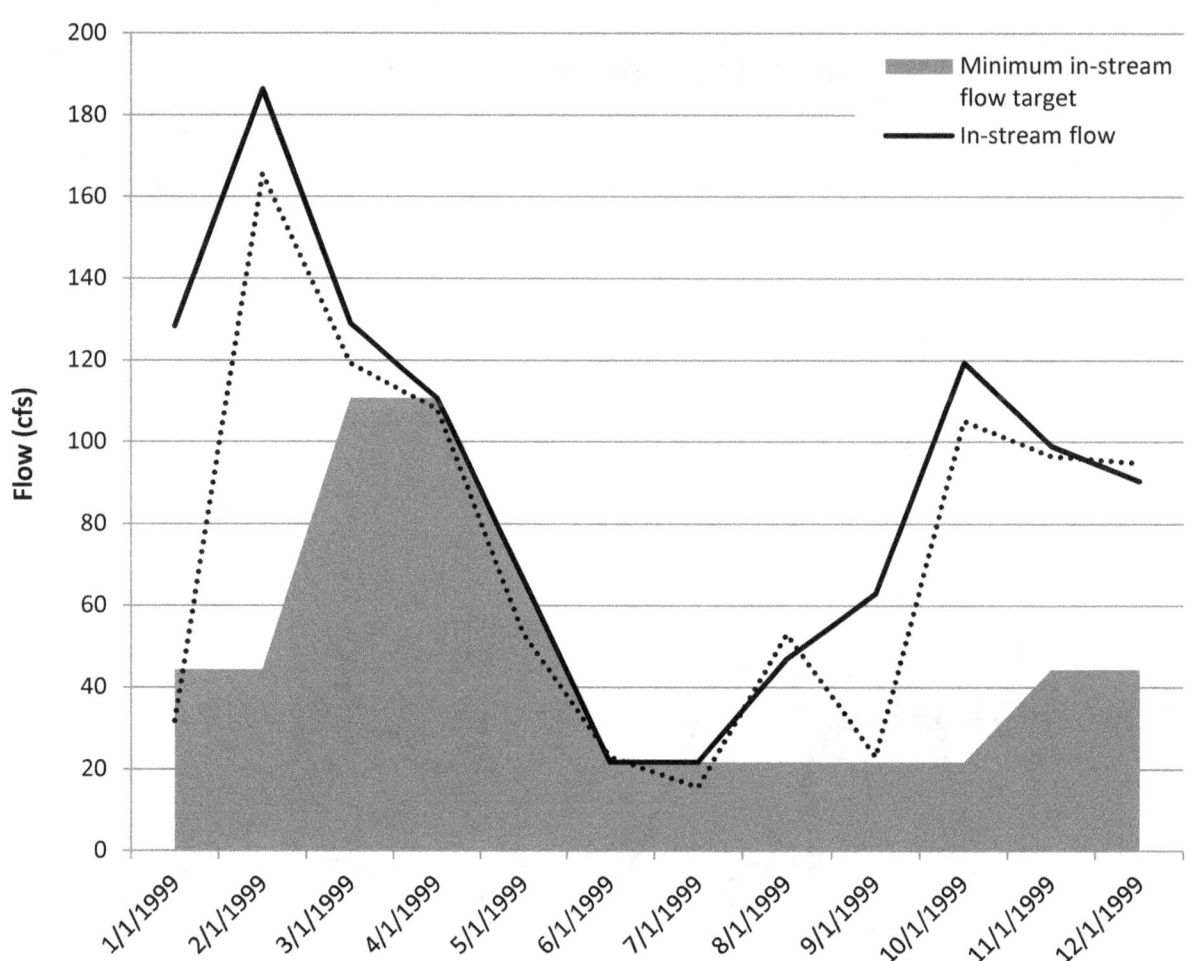

3.2   Danvers and Middleton, Massachusetts

The towns of Danvers and Middleton, Massachusetts (MA) were selected as a case study because they are a pilot project for the Massachusetts Sustainable Water Management Initiative (SWMI)[33] and because they are located in the Ipswich River Basin (IRB) which experiences low and no flow events in the late summer. The Ipswich River is the primary source of water for these towns. The site-specific data used in this case study are from the U.S. Geological Survey (USGS) Hydrological Simulation Program-Fortran (HSPF) simulation model, SWMI Framework, state databases and websites, and the towns' websites. *Please see Appendix A for details on input data values, sources/references and assumptions. We have excluded some detail for readability and to keep focus on effects of scenario specifications on changes in management practices and costs.*

Please note that we were unable to coordinate with Danvers Water Division to corroborate data, assumptions and interpretation of the results. In addition, the SWMI Framework does not specify quantitative, minimum in-stream flow criteria for the basin in which Danvers and Middleton are located. In order to run the model, we used the first basin category in the SWMI Framework with quantitative criteria (i.e., least stringent of the quantitative criteria) to allow a hypothetical case study and to be analyzed to demonstrate the application of WMOST for municipalities rather than an entire watershed.

The Ipswich River Basin is 155 square miles, approximately a HUC-10 watershed. The town of Middleton is entirely within the IRB while only 28% of Danvers drains to the Ipswich River (*Exhibit* 7). The populations of Danvers and Middleton were 26,493 and 8,987 respectively, in 2010.

Exhibit 7. Danvers and Middleton, MA in the IRB

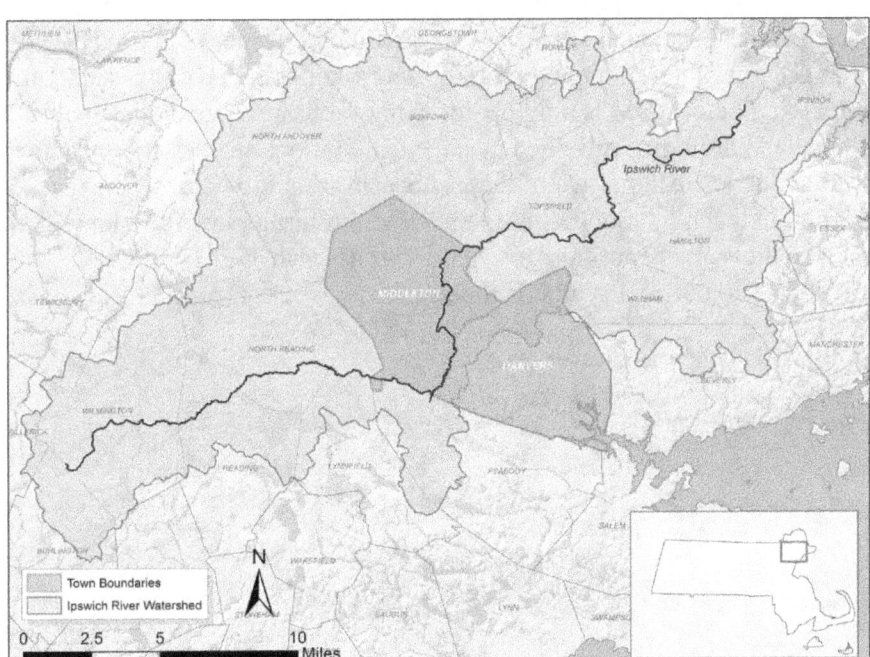

---

[33] http://www.mass.gov/eea/air-water-climate-change/preserving-water-resources/sustainable-water-management/

Danvers maintains the sole water treatment plant and manages the distribution of water to Danvers and Middleton. Middleton purchases all of its water from Danvers. Three surface water sources and two groundwater wells serve as the source of water supply. Middleton Pond serves as the primary supply and is supplemented with water from Emerson Brook Reservoir in Middleton, and Swan Pond in North Reading during months of high demand.

In Danvers, wastewater is 99% sewered and exported to the South Essex Sewer District which discharges outside of the IRB. Middleton's wastewater primarily discharges to septic systems except for three properties.

Total water demand for Danvers and Middleton system (DM) has decreased since the 1990s due to demand and other management efforts. However, DM expects to need 0.27 MGD in their new withdrawal permit beyond the SWMI baseline withdrawals. SWMI baseline withdrawals for DM pilot study in the SWMI Phase I Report are calculated as 2005 demand plus a growth factor of 8%. Withdrawals beyond this amount require various levels of minimization of withdrawals and/or mitigation of withdrawal impacts. Depending on the basin, in-stream flow criteria may apply. While Danvers is in a basin that does not have quantitative flow criteria, we used the first category of basins with flow criteria in order to run the model. Therefore, with a need for additional water for human demand and specifying quantitative in-stream flow targets, we expected that the management solution would require new management practices and, therefore, provide an illustrative example of the application of WMOST.

3.2.1 Model Setup

Part of the challenge in jointly modeling human and natural systems is that they often do not overlap. DM is part of 18 subbasins of the IRB, as illustrated in *Exhibit 3*. The USGS developed a detailed HSPF model of the IRB (Zarriello and Ries, 2000 and Zarriello, 2002) and which we used to obtain the data necessary to model the DM subbasins. These data include the area of hydrologic response units (HRUs)[34] within subbasins, HRU runoff and recharges rates, pumping rates, groundwater storage volumes and other (see Appendix A). As shown in *Exhibit 8* below, DM is only part of many subbasins. We modeled all subbasins that overlap with DM for the hydrology but limited management options such as stormwater practices to land areas within DM boundaries. All but two of these subbasins drain to consecutive reaches of the Ipswich River with a pour point in reach 37. Subbasins 45 and 46 drain further downstream with a pour point in reach 46 which drains to the main stem of the Ipswich River at reach 47. Therefore, we aggregated outflow from reach 37 and 46 to derive a synthetic time series to use as 'measured flow' and compare against WMOST modeled flow.

---

[34] HRUs are areas of similar hydrology due to similar characteristics such as land use, soil and/or slope.

Exhibit 8. Subbasins and Reaches of the IRB and Danvers and Middleton Town Boundaries

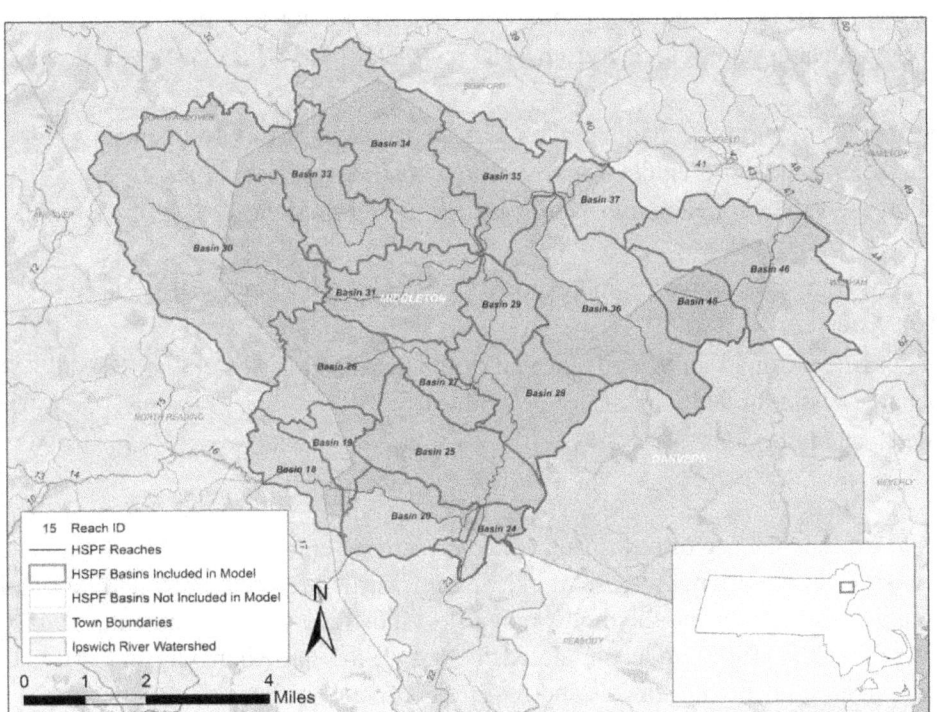

In the IRB HSPF model HRUs are defined as a combination of land use and surficial geology with a total of 11 HRUs in the DM subbasins. It is important to note that in the HSPF model, wetlands were not simulated as land use but rather as stream reaches that were "placed" between the runoff from the HRUs and the actual stream reach. Because wetlands change in area over which they have standing water, the hydrology of wetlands changes significantly over the simulation period. For example, during the spring and early summer, there may be significant evapotranspiration from wetlands. As they become more dry and shrink in area, the amount of evapotranspiration decreases. This change in area and hydrology was simulated using reaches that could be programmed to change area based on depth of water and geometry of the channel. Because wetlands were not modeled as HRUs, there are no runoff and recharge rates available for them. As a result, we were not able to include wetlands in the WMOST model at this time and would require consultation with USGS on the most appropriate way to translate their modeling of wetlands into the WMOST modeling structure and/or develop new functionality in WMOST.

3.2.2 Simulation

The first modeling step was to determine the accuracy of using data from the IRB HSPF model in WMOST. Therefore, we used the hydrology and pumping data from HSPF and compared WMOST modeled in-stream flow with the HSPF synthetic gage flow. Comparing with the synthetic gage flow was necessary due to subbasin with DM that did not drain to consecutive stream reaches. The time period of simulation was limited to the available surface water and groundwater pumping data in the HSPF model which covered the years from 1989 to 1993. The simulation run used the following data:

- Land areas, runoff rates and recharge rates for 11 HRUs for 1990
- Surface water:

- External inflow to the study area calculated as sum of inflow from upstream subbasins
  - Reservoir/Surface water storage: three ponds with usable storage
- Groundwater minimum, maximum and initial storage as well as recession coefficient
- Surface water and groundwater pumping data from 1989 to 1993
- Human demand:
  - Disaggregated HSPF pumping data based on MA Department of Environmental Protection (DEP) Annual Statistical Report (ASR) data into five user types
  - Consumptive use values for each user were based on literature data
- Wastewater:
  - Assumed all of Danvers is sewered and exported to South Essex Sewer District (i.e., interbasin transfer)
  - Assume all of Middleton is septic and since all of the town area is within the IRB, all septic discharge was assumed to recharge groundwater
- Exclusion of all management options
- No in-stream flow targets

Although there are options to exclude the use of new management practices, currently there is no capability to prevent the model from optimizing the use of existing infrastructure. For example, DM has both surface water and groundwater sources and the model may select a different timing and amount of withdrawals from each source than the HSPF model simulated. To ensure the same behavior in WMOST as in the HSPF model, we input all human withdrawals and returns as private surface and groundwater withdrawals and discharges under the surface water and groundwater input sheets as follows:

- Surface withdrawal: HSPF surface water withdrawals

- Groundwater withdrawal: HSPF groundwater withdrawals and loss of groundwater due to infiltration to wastewater collection system

- Groundwater discharge: Unaccounted for water calculated as percentage of total pumping and septic returns from Danvers adjusted for consumptive use

Using the above data in the specified configuration, WMOST reproduced HSPF daily in-stream flow over the five years with a Nash-Sutcliffe Efficiency (NSE) of 0.93.[35] This value is an overall assessment of model fit for the entire time period (see *Exhibit 9* below). In the exhibit, HSPF flows were used to represent "measured" flows which are compared against WMOST modeled in-stream flows. WMOST modeled summer low-flows higher than HSPF flows. Low flows are an important element for the DM case study because of the low flow issues in the IRB. A potential source of difference is the evapotranspiration from wetlands. As explained in the model setup, the IRB HSPF model did not model wetlands as a land area and, therefore, could not be included in the WMOST model.

In future refinements of this case study,[36] it would be beneficial to consult with USGS on the most appropriate way to represent the effects of wetlands in WMOST and other factors that may be contributing to higher than measured low-flows. For example, it may be necessary to have two

---

[35] The NSE ranges from 0 to 1. A value of 0 indicates that the model estimates values only as well as the average of the measured data. A value of 1 indicates a perfect match to the measured data.

[36] We provide recommendations throughout the case study but all recommendations are summarized at the end of the case study section.

groundwater storage components in the model - one to represent interflow which discharges more quickly in response to recharge from storm events and another to represent active groundwater flow which discharges at a slower and steadier rate. Alternatively, or in addition, it may be necessary to re-calibrate the groundwater recession coefficient from the HSPF value to a value more representative for the lumped spatial characteristics of WMOST. However, with respect to this case study, the low flows modeled are lower than the quantitative, minimum in-stream flow targets. Therefore, we increased the minimum flow targets by the percent difference between HSPF and WMOST flows. We were able to proceed with optimization runs given that the model will need to select management practices that will provide additional in-stream flow. As such the case study will provide insight into which management practices are most cost-effective for increasing in-stream flow while meeting anticipated increase in demand.

Exhibit 9. Comparison of In-Stream Flows

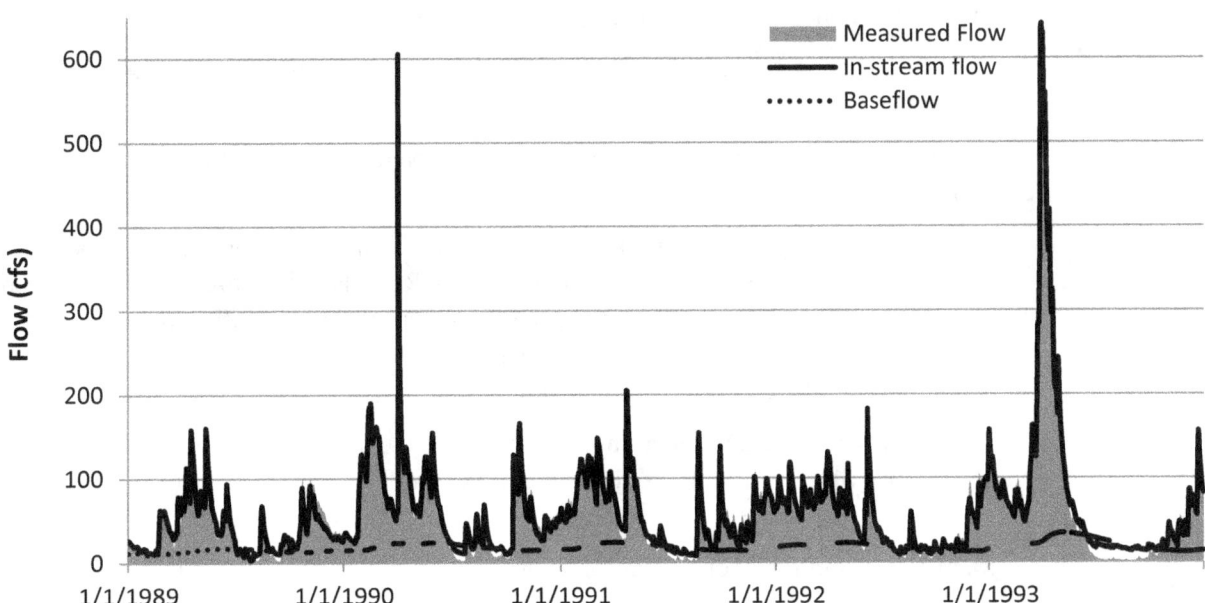

Note: In some cases baseflow is higher than modeled in-stream flow. In-stream flow receives baseflow but also has withdrawals; therefore, final flow in the stream may be lower than baseflow.

3.2.3 Optimization

The optimization was run to simultaneously meet two management goals at least cost: 1) projected increase in human demand for DM based on baseline SWMI demand plus additionally requested withdrawals by DM[37] and 2) quantitative, minimum in-stream flow targets. We assumed a 20-year planning period based on water withdrawal permit lifetimes and projected build-out land use based on multiple data sources (see Appendix A for input data sources and values).

The following management practices were available for meeting the goals:

1. Meeting human demand by:

---

[37] Note that the projected demand of 3.72 MGD is lower than the 1993 pumping of 4.56 MGD in the HSPF model.

    a. Increasing withdrawals from surface and/or ground sources,

    b. Purchasing interbasin transfer of water from the Massachusetts Water Resources Authority (MWRA),

    c. Reducing demand via increased pricing of water services, and

    d. Reducing demand by providing rebates for water efficient appliances.

2. Meeting minimum in-stream flow criteria by:

    a. Changing the rate and timing of pumping from surface and ground sources to alleviate summer low flows,

    b. Reducing withdrawals by:

        i. Meeting human demand using options 1b through 1d,

        ii. Constructing a water reuse facility and nonpotable distribution system to reuse water,

    c. Increasing recharge to groundwater and, therefore, potentially increasing late summer baseflow[38] by:

        i. Implementing infiltration-based stormwater management practices by allocating urban HRU areas to one of six "managed" HRU sets each representing a different stormwater management practice (i.e., bioretention basins, infiltration basins, or horizontal wetlands sized to manage a 0.6 inch storm event; bioretention basins, infiltration basins, or horizontal wetlands sized to manage a 2 inch storm event),

        ii. Constructing an aquifer storage and recharge (ASR) facility to recharge groundwater during flows above the in-stream flow criteria or from reservoirs/surface water storage,

        iii. Conserving forest or undeveloped open land,

    d. Increasing surface storage capacity (e.g., size of storage tank, reservoir, pond),

    e. Constructing a wastewater treatment plant and discharging to the Ipswich River rather than exporting out of the basin, and

    f. Reducing the infiltration and inflow (I/I) of groundwater into the sewer collection system.

Currently, WMOST is configured such that I/I into the sewer system flows to wastewater treatment plant. In DM, all sewered wastewater is exported via interbasin transfer. Therefore, specifying an I/I rate would create flow to a wastewater treatment plant rather than to interbasin transfer. In some of the scenarios (Scenario 3 and 4), constructing a wastewater treatment plant is selected. In these scenarios, the I/I rate is specified and its repair is available as a management practice. The modeling capability to specify I/I for interbasin transfer is noted as a desired enhancement in *Section 3.2.4* where we summarize future recommendations for case study input data and model capabilities.

---

[38] The timing and amount of recharge reaching a stream as baseflow depends on site specific parameters such as whether the groundwater beneath the recharge area flows to the stream, the distance between recharge area and stream, and whether there are any withdrawal wells between the stream and recharge area.

For the first optimization scenario, we set all of the above management practices to be available to meet management goals of projected demand and minimum in-stream flow targets. *Exhibit 10* shows the non-zero values from the results table.

While the objective of this case study is to minimize costs to meet management goals, it is interesting to note that the total annual cost is lower than the water and wastewater revenue. The O&M costs for water and wastewater were derived from online town budgets for the water and wastewater departments. The total cost shown by the model approximately matches the sum of costs specified in the budgets (approximately $5 million for water and $5 million for wastewater for both towns). However, these costs may not include all costs incurred by the towns. For example, the comprehensive costs of providing water services may include operation of source pumps, treatment, operation of the distribution system, capital improvement, source protection, and administrative costs such as billing customers. We could not determine additional costs the towns may incur that are not presented in the water department budgets. Therefore, future refinement of this case study would require collaboration with the towns to understand their comprehensive costs.

With respect to minimizing costs to meet human and in-stream flow demand, the model selected to:

- Implement pricing change to the maximum extent allowed (20% increase over the 20-year planning period),

- Provide rebates for water efficient appliances to reduce demand at the maximum available 0.6 MGD possible, and

- Repair leakage in the potable distribution system to the maximum extent possible (fixing 99% of the leaks was set as practical limit).

The water treatment plant capacity is sufficient; therefore, no additional capacity is necessary. The only costs for the water treatment plant are those associated with annual operations. For wastewater, interbasin transfer continues to be the sole service provider.

Exhibit 10. Base Scenario Results

Total Annual Cost	$13.4	million
Water Revenue	$10.2	million
Wastewater Revenue	$10.3	million

MANAGEMENT PRACTICES	UNITS	Number of Units	Total Annual Sub-Costs (incl. O&M)
Consumer Rate Change	%	20	$3,846
Direct Demand Reduction	MGD	0.60	$255,701
Additional WTP Capacity	MGD	0.00	$6,721,130
Potable Distribution System Repair	% of Leaks	99	$138,179
Additional IBT - Wastewater	MGD	0.00	$6,271,870

These results suggest that the more cost effective management practices are demand reduction via pricing changes, direct demand management through providing rebates, distribution system repair, and continuing with local water withdrawal and interbasin transfer of wastewater.

Implementation of these selected management practices increases the modeled in-stream flows as shown in *Exhibit 11*, below. The modeled flows are greater than the specified streamflow criteria for all days in the five-year modeling period as shown in *Exhibit 12*.

Exhibit 11. Base Scenario – Modeled and Measured In-Stream Flows

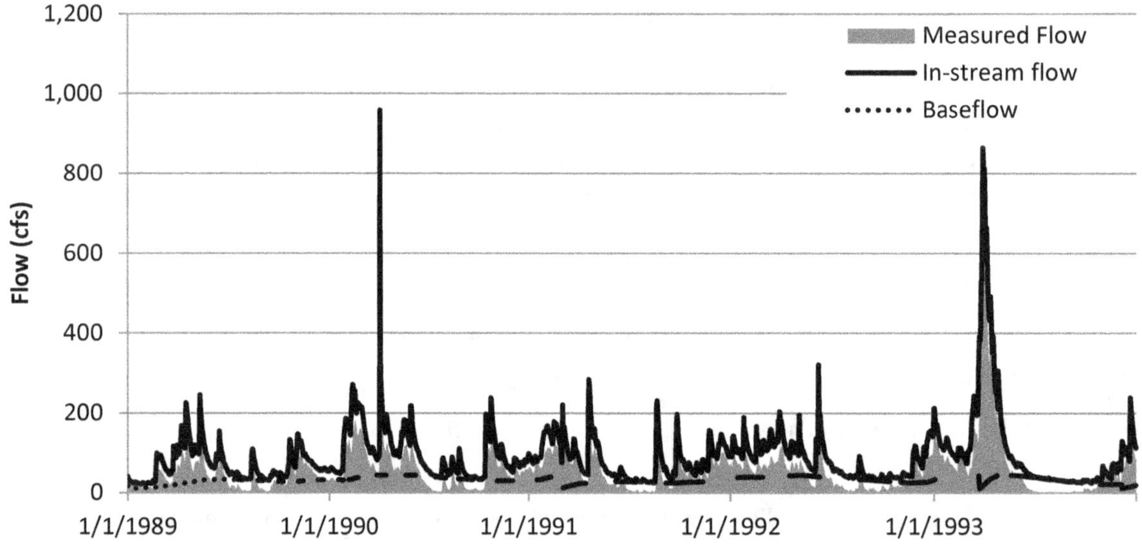

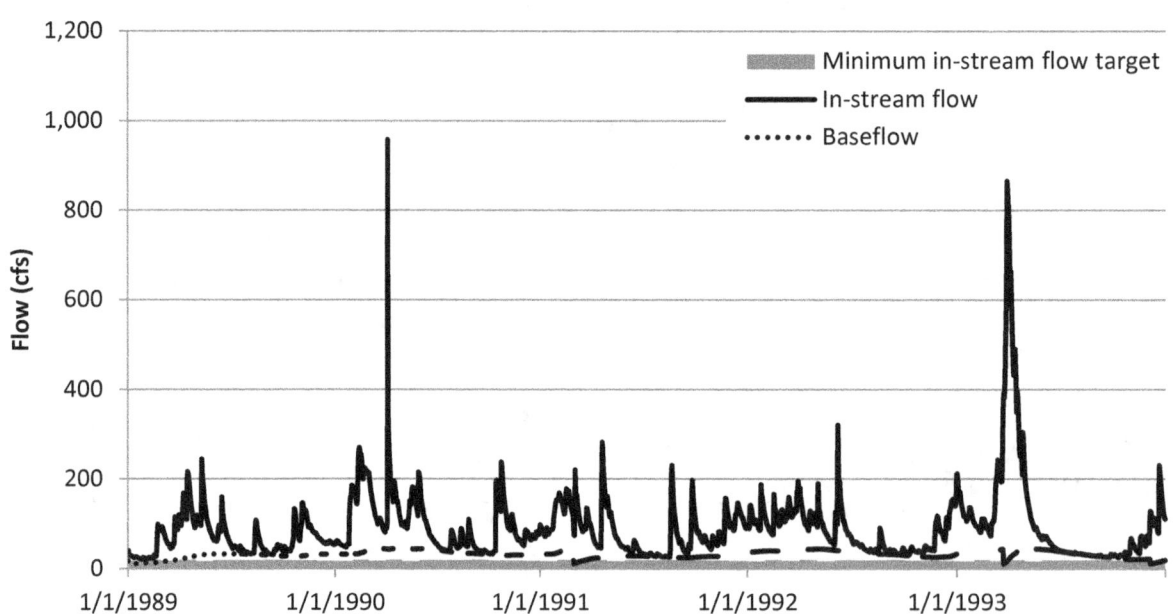

Note: Baseflow may be higher than modeled in-stream flow. In-stream flow receives baseflow but also has withdrawals; therefore, final flow in the stream may be lower than baseflow.

Exhibit 12. Base Scenario - Modeled and Target In-Stream Flows

In-stream flow criteria are primarily exceeded in the late summer after a season of reduced precipitation, increased evapotranspiration and increased human use (see *Exhibit 5* in *Section 3.1*). A significant portion of human use during the growing season (approximately May to September) is attributable to outdoor watering. The efficiency of outdoor watering may be increased via moisture sensors and installation of low water need/drought resistant landscaping. In addition, streamflow-triggered watering limits or bans can reduce outdoor watering. Although the towns already have efforts underway to reduce outdoor watering, we wanted to examine the effect of investing in additional practices as mentioned above. In this scenario, we calculated a reduced summer outdoor water use as follows:

1. Calculate the average daily demand for the summer months (May through September) when outdoor watering is expected and for the winter months (December through February) when outdoor watering is not expected,

2. Determine the difference between the two daily rates and assume it represents outdoor water use,

3. Take 50% of the difference, thereby assuming that outdoor watering use could be reduced by 50%,

4. Subtract this amount from daily summer demand.

We also calculated new consumptive use percentages assuming that all reduction in demand is 100 percent consumptive.

Rerunning the model with these specifications reduced the total annual management cost, as may be expected. The same management practices were selected as in the Base Scenario but with less demand less water was provided and total cost decreased. It is important to note that both water and wastewater revenue decreased. Most towns charge wastewater services based on metered water flow as is the case in DM and in the model. While outdoor watering does not discharge wastewater, most towns' wastewater services may be dependent on the total annual revenue according to current practices of charging customers. Therefore, water conservation programs must prepare both the water and wastewater departments for reduced revenues and potential need for rate increases that reflect the cost of services provided and received flow, respectively. Since many towns are making allowance for separate outdoor water meters and/or adjustment of summer wastewater charges based on winter water flow rates and water conservation is a cost effective management practice as also shown in the Base Scenario, issues of sustainable water and wastewater rates may be important to address.

Exhibit 13. Reduced Summer Water Use Scenario Results

Total Cost		$12.8	million
Water Revenue		$9.5	million
Wastewater Revenue		$9.6	million

MANAGEMENT PRACTICES	UNITS	Number of Units	Total Annual Sub-Costs (incl. O&M)
Consumer Rate Change	%	20	$3,846
Direct Demand Reduction	MGD	0.60	$255,701
Additional WTP Capacity	MGD	0.00	$6,239,750
Potable Distribution System Repair	% of Leaks	99	$138,179
Additional IBT - Wastewater	MGD	0.00	$6,198,780

For model enhancements, we recommend adding a management practice that reduces summer demand based on a user specified streamflow threshold. This modeling capability would represent programs some towns implement in the summer to limit outdoor watering during low flows. In addition, adding a direct demand management practice that allows for reductions in specific months would also provide additional flexibility for modeling outdoor water use management practices.

Scenario 3: Trade-off Curve–In-Stream Flow and Costs

To evaluate trade-offs between in-stream flows and costs, we ran the case study five times with the base scenario while increasing the in-stream flow criteria each time. We chose to increase the base criteria by 25, 50, 75 and 100 percent (i.e., set the minimum in-stream flows to 125%, 150%, 175%, and 200% of base values). These percent increases translate to the flow criteria shown in *Exhibit 14*.

This series of runs provides insight into the: 1) trade-off between total cost and increasing in-stream flows beyond the minimum criteria, 2) the relative cost effectiveness of practices that were not selected in previous scenarios, and 3) additional information about potential management practices that may be necessary to meet minimum in-stream flow once the model is better able to reproduce low-flows. The results of these runs are shown below in *Exhibit 15* and the trade-off curve between total cost and in-stream flow is shown in *Exhibit 16*.

Exhibit 14. In-Stream Flow Criteria

	In-Stream Flow (cfs)					
	Base (100%)	*125%*	*150%*	*175%*	*200%*	*Average HSPF Flow (1989-1993)*
January	16.56	20.69	24.83	28.97	33.11	59.95
February	19.10	23.88	28.65	33.43	38.20	80.32
March	17.28	21.59	25.91	30.23	34.55	90.23
April	19.46	24.33	29.19	34.06	38.92	132.82
May	15.98	19.98	23.97	27.97	31.96	66.01
June	18.51	23.13	27.76	32.38	37.01	42.06
July	18.46	23.07	27.68	32.30	36.91	9.41
August	18.76	23.44	28.13	32.82	37.51	19.92
September	18.85	23.56	28.28	32.99	37.70	14.85
October	17.52	21.89	26.27	30.65	35.03	37.37
November	17.52	21.89	26.27	30.65	35.03	54.15
December	17.09	21.36	25.63	29.90	34.17	64.77

Exhibit 15.

Results for Increasing In-Stream Flow Criteria

MANAGEMENT PRACTICES	UNITS	In-stream Flow Criteria							
		125%		150%		175%		200%	
		Number of Units	Total Annual Sub-Costs (incl. O&M)	Number of Units	Total Annual Sub-Costs (incl. O&M)	Number of Units	Total Annual Sub-Costs (incl. O&M)	Number of Units	Total Annual Sub-Costs (incl. O&M)
Consumer Rate Change	%	20	$3,846	20	$3,846	20	$3,846	20	$3,846
Direct Demand Reduction	MGD	0.60	$255,701	0.60	$255,701	0.60	$255,701	0.60	$255,701
Additional WTP Capacity	MGD	0.00	$6,721,130	0.00	$6,721,130	0.00	$6,721,130	0.00	$6,721,130
Potable Distribution System Repair	% of Leaks	99	$138,179	99	$138,179	99	$138,179	99	$138,179
Additional IBT - Wastewater	MGD	0.00	$6,271,870	0.00	$6,255,920	0.00	$6,259,650	0.00	$6,208,070
Infiltration basin, 0.6"	Acres			1,255	$570,206	1,255	$570,206	1,255	$570,206
Additional WWTP Capacity	MGD			0.75	$706,592	0.75	$701,921	0.75	$766,399
Additional ASR Capacity	MGD			0.71	$534,736	5.04	$3,815,700	9.49	$7,853,070
Additional WRF Capacity	MGD			0.06	$44,680	0.01	$8,871	0.20	$140,575
Total Cost	millions		$13.4		$15.2		$18.5		$22.7
Water Revenue	millions		$10.2		$10.2		$10.2		$10.2
Wastewater Revenue	millions		$10.3		$10.3		$10.3		$10.3

The 25 percent increase in in-stream flow criteria produced the same result as the base scenario. This result suggests that the base scenario was not entirely limited by in-stream flow and flexibility remained in the system to meet a higher in-stream flow with the same set of management practices..

Achieving a 50 percent increase in minimum in-stream flows, however, requires, additional management practices, including:

- stormwater management using infiltration basins sized for 0.6-inch storm event on commercial land use with sand and gravel surficial geology,

- local wastewater treatment plant,

- water reuse facility (WRF) to supply the ASR facility (see below) with additionally treated wastewater from the local wastewater treatment plant, and

- aquifer storage and recharge (ASR) facility that utilizes water from the surface water, reservoir and WRF.

Exhibit 16. Trade-off Curve Between Increasing In-stream Flow and Total Cost

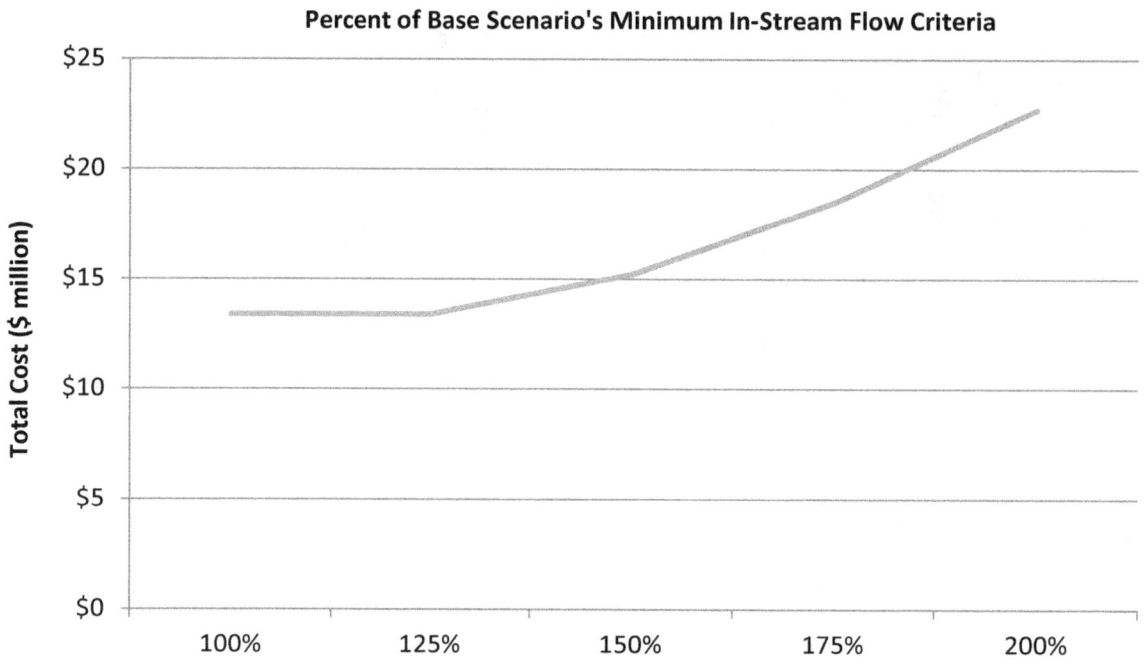

- Interestingly, the wastewater treatment plant is constructed to a maximum capacity of 0.75 MGD but used variably among the scenarios as reflected in the total cost which includes costs for O&M for the plant. Interbasin transfer of wastewater is relatively less expensive than local wastewater treatment similar to the interbasin transfer of water because of economies of scale for the larger system from which these services are bought. Therefore, although the model selects to build a wastewater treatment plant, it only uses it during critical times when additional discharge is necessary to meet in-stream flow. This suggests that not only is there a need to reduce demand and withdrawals but also a need to increase discharge and recharge in the basin and river.

However, treatment plants must have a predictable flow, and therefore, this solution would not be practical. As a result, we ran another scenario where interbasin transfer of wastewater was excluded and in-stream flow criteria were set at 200% of the base scenario (see Scenario 5 below).

- In continuing to examine the results of increasing in-stream flow criteria, we also note that ASR is selected at an increasing capacity. Such high recharge capacities may not be feasible within the land area and aquifer storage of DM. Therefore, it would be necessary to determine the feasibility of the maximum amount of ASR and limit this management option to the feasible capacity.

- Finally, we note the use of infiltration basins for a 0.6-inch design storm on 1,255 acres of commercial land on sand and gravel. Commercial land has the highest percent impervious cover while sand and gravel had the highest infiltration rate. These characteristics likely resulted in this HRU being the most cost effective for stormwater management. No other HRUs were selected for stormwater management.

- Given that infiltration basins and ASR system both recharge groundwater, it is interesting to note that both management practices were selected. It is likely that infiltration basins on commercial land over sand and gravel are more cost effective than ASR but the same does not hold true for other HRUs. With future refinements, if a feasibility limit is set for ASR, stormwater management for other HRUs may become relatively more cost effective.

Scenario 4: Exclusion of Interbasin Transfer of Wastewater and Double In-Stream Flow Criteria

Both interbasin transfer of wastewater and local wastewater treatment were selected as management options in the high in-stream flow criteria scenarios. To evaluate the sensitivity of the selected practices to the availability of certain practices, we re-ran the scenario requiring double the base scenario in-stream flow but excluding interbasin transfer of wastewater. As may be expected, total costs increased but otherwise the selection of management decisions were the same with slight differences in the maximum capacity of ASR and WRF. It is possible that a town may elect to construct a treatment plant for the minimum expected need for meeting in-stream flow and continue with interbasin transfer for additional wastewater needs. In such a case, an additional scenario could be run where interbasin transfer is limited to the total need minus the 0.75 MGD of local wastewater capacity selected in earlier scenarios. With the inclusion of a wastewater treatment plant as in Scenario 3, I/I could be specified and the model selected to repair I/I to the maximum extent possible. This result suggests that repair of groundwater flow into sewer collection system is less expensive than treating a larger volume of wastewater and/or helps retain groundwater in the aquifer for baseflow and is cost effective for meeting in-stream flow criteria.

Exhibit 17. Results for Scenario with Exclusion of Interbasin Transfer of Wastewater with Double In-Stream Flow Criteria

Total Cost	$28.2	million
Water Revenue	$10.2	million
Wastewater Revenue	$10.3	million

MANAGEMENT PRACTICES	UNITS	Number of Units	Total Annual Sub-Costs (incl. O&M)
Consumer Rate Change	%	20	$3,846
Direct Demand Reduction	MGD	0.60	$255,701
Additional WTP Capacity	MGD	0.00	$6,721,130
Potable Distribution System Repair	% of Leaks	99	$138,179
Additional IBT - Wastewater	MGD	NA	NA
Additional WWTP Capacity	MGD	5.52	$12,938,300
Infiltration Repair	% of Leaks	99	$38,337
Infiltration basin, 0.6"	Acres	1,255	$570,206
Additional ASR Capacity	MGD	9.27	$7,231,130
Additional WRF Capacity	MGD	0.44	$344,822

Scenario 5: Sensitivity of Solutions to the Capital Cost of Interbasin Transfer of Water

The purchase of MWRA water was not selected as a management practice in any of the previous scenarios. The source of the capital cost for purchasing MWRA water cited the value as an order-of-magnitude estimate. To determine the sensitivity of suggested management practices to the capital cost of purchasing MWRA water, we ran multiple scenarios decreasing this capital cost. The purchase of MWRA water was not selected until the capital cost was reduced to less than 25% of the initial estimate. *Exhibit 18 below shows results for capital costs that are 20% of the initial estimate.*

In varying this capital cost, MWRA water was always either selected at full availability of 0.27 MGD or not at all. This suggests that once the capital cost is reduced enough to make the practice cost effective, the O&M or purchase cost of incremental volumes of water is less than that of the local water treatment plant. This is further confirmed by the fact that the solutions for the other management practices do not change except for a decrease in the amount of water withdrawn and provided by the local water treatment plant as reflected in the lower cost for the water treatment plant. Similar to wastewater treatment, economies of scale result in lower O&M costs for MWRA than the smaller local water treatment plant ($3,803/MG for MWRA compared to $5,314/MG for town produced water).

Although these solutions suggest that MWRA water connection is not cost effective until the capital cost is reduced by more than 75%, this conclusion is based on scenarios that include a potentially infeasible quantity of ASR. Therefore, once the maximum feasible ASR quantity is determined and limited, this sensitivity analysis for the capital cost of MWRA water may be repeated to determine the capital cost at which the management practice may be cost effective. Finally, it may be possible to finance or amortize the MWRA capital cost over a longer time period than the planning horizon of 20 years. A longer amortization period would lead to a lower annual payment or cost.

These sensitivity runs demonstrate the importance of accurate input data. In addition, if the uncertainty or range of values for an input is known, it provides an approach to determine whether that management practice and the overall solution will remain constant over that range of values.

Exhibit 18. Results for Reducing the Capital Cost of Interbasin Transfer of Water

Total Cost	$22.6	million
Water Revenue	$10.2	million
Wastewater Revenue	$10.3	million

MANAGEMENT PRACTICES	UNITS	Number of Units	Total Annual Sub-Costs (incl. O&M)
Consumer Rate Change	%	20	$3,846
Direct Demand Reduction	MGD	0.60	$255,701
Additional WTP Capacity	MGD	0.00	$6,197,090
Potable Distribution System Repair	% of Leaks	99	$138,179
Infiltration basin, 0.6"	Acres	1,255	$570,206
Additional WWTP Capacity	MGD	0.75	$766,399
Additional IBT - Potable	MGD	0.27	$502,869
Additional IBT - Wastewater	MGD	0.00	$6,208,070
Additional ASR Capacity	MGD	9.49	$7,853,070
Additional WRF Capacity	MGD	0.20	$140,575

Conclusions

Over all scenario runs, WMOST suggests several of the same management practices for meeting human demand and in-stream flow criteria. These management practices are likely to be most cost effective, and include:

- Demand management through pricing changes,

- Direct demand reduction via rebates,

- Repair of leakage from the potable distribution system, and

- Repair of infiltration into the sewer collection system.

Several management practices – local wastewater treatment and discharge, stormwater management, ASR, and WRF – were selected as in-stream flow criteria were increased. Some practices such as ASR were selected at capacities that are likely not feasible. Therefore, further input data based on a feasibility study[39] and additional modeling capabilities are necessary to limit some options to remain within realistic limits (e.g., maximum limit on addition ASR capacity).

---

[39] ASR is practiced in the Western and Southeast U.S. However, we did not find any ASR wells nor feasibility studies for ASR in the Northeast (http://water.usgs.gov/ogw/artificial_recharge.html, http://water.epa.gov/type/groundwater/uic/aquiferrecharge.cfm#inventory).

The DM case study example shows that, by running various scenarios, it is possible to identify the most promising and cost effective management options for meeting management goals, and to assess the extent to which these options remain cost effective under different sets of assumptions. However, the case study also illustrates the importance of the input data and understanding the system dynamics to appropriately interpret results and to conduct additional scenario runs, as needed, to ensure that relationships between costs and the effect of management practices are accurately determined.

3.2.4 Refinements for Input Data and WMOST Capabilities

Based on this case study, the following enhancements would be useful in future versions of WMOST and refinements of the DM case study:

- Module to help calibrate simulation for a few, key parameters that may be least accessible such as initial, maximum, and minimum groundwater storage and the groundwater recession coefficient.

- Module specifically formulated for simulation such that all decisions are excluded from optimization. Currently the use cannot exclude the optimization of existing infrastructure operations (e.g., the amount of surface water versus groundwater that is used to meet demand).

- More options for interbasin transfer. For example, I/I could not be represented when only interbasin transfer was used for wastewater services. However, the collection system in a town may still have I/I into pipes that connect to the service provider outside of the basin. Therefore, an option to have I/I and its repair represented even when all wastewater is transferred out of basin would make the model more flexible and able to accurately represent more systems.

- For revenue calculations, additional input data would be needed but would make the calculation more accurate if the user had the ability to specify fixed- and flow-based rates per water user type (e.g., different rates for commercial and residential).

- In-stream flow triggered reduction in demand to represent outdoor water limits or bans that may be implemented based on in-stream flow. For example, specify the reduction in demand when in-stream flow falls below a user-specified threshold.

- Direct demand management practice for which reduction in demand can be specified by month.

- Sensitivity module to determine the point at which different management practices may be selected. This is especially important for costs. For example, a feature that would allow user-selected or all costs to be varied by +/-10% to determine effects on the selected combination of management practices.

- Maximum feasibility determination for ASR capacity for DM.

4. Appendix A. Danvers Middleton Case Study Input Data

Description	Value Used in Model	Original Values	Data Source
LAND USE			
Number of land use sets (HRUs)	11	11	HSPF
Stormwater Management Sets	6	6	6 BMPs: Bioretention, infiltration basin and horizontal wetland designed for the 0.6" and 2.0" storm; Original HSPF runoff and recharge rates were modified based on EPA Region 1 Stormwater DSS model for these BMPs
Case study area	23,810 acres	HSPF subbasins that overlap with DM	HSPF subbasin shapefile
Scenario land area	see model for individual values	See User guide for LU calculations	HSPF shapefiles, MassGIS town, protected openspace, zoning, 2005 land use layers
Minimum area for each land use	see model for individual values	See User guide for LU calculations	HSPF shapefiles, MassGIS town, protected openspace, zoning, 2005 land use layers
Maximum area for each land use	see model for individual values	See User guide for LU calculations	HSPF shapefiles, MassGIS town, protected openspace, zoning, 2005 land use layers
Capital cost to conserve land use	$187,408/acre	Average of 2 forested lots: Mls #: 71204964, Mls #: 71156281	http://www.verani.com/real-estate/Middleton/MA/land
O&M cost to conserve land use	$1874.08/acre	1% of capital costs	BPJ
Stormwater Management			
Capital installation cost	see case study model for individual costs	Cost per Acre for each BMP	Charles River Watershed Association/Horsely-Witten Group
O&M cost	see case study model for individual costs	Cost per Acre for each BMP	Charles River Watershed Association/Horsely-Witten Group
RUNOFF AND RECHARGE			
Recharge rates for each original or "baseline" land use	in/day	see case study model for individual values	HSPF - interflow plus recharge
Runoff rates for each original or "baseline" land use	in/day	see case study model for individual values	HSPF - runoff
Recharge rates for each "managed" land use	in/day	see case study model for individual values	6 BMPs: Bioretention, infiltration basin and horizontal wetland designed for the 0.6" and 2.0" storm; Original HSPF runoff and recharge rates were modified based on EPA Region 1 Stormwater DSS model for these BMPs
Runoff rates for each "managed" land use	in/day	see case study model for individual values	6 BMPs: Bioretention, infiltration basin and horizontal wetland designed for the 0.6" and 2.0" storm; Original HSPF runoff and recharge rates were modified based on EPA Region 1 Stormwater DSS model for these BMPs
WATER DEMAND			
Demand for each user for each day	see case study model for individual values	see case study model for individual values	Optimization: SWMI 2005 baseline + 8% growth factor + additionally requested withdrawal by Danvers (SWMI Phase I Report). User types: Average over 2010-2012 from DEP ASRs: 6% UAW, 58% Residential, 26% Commercial, <1% Agricultural, <1% Industrial, 10% Municipal
Unaccounted-for-water (i.e., leakage from potable water distribution system)	6%	6%	MA DEP ASRs 2010-2012
Percent consumptive use for each water user for each month	see case study model for individual values	see case study model for individual values	Based on data from Amy Vickers (2002) Handbook of Water Use and Conservation
Nonpotable water			
Maximum percent demand that can be met by nonpotable water for each user	see case study model for individual values	see case study model for individual values	Based on data from Amy Vickers (2002) Handbook of Water Use and Conservation
Percent consumptive use for nonpotable water for each user for each month	see case study model for individual values	see case study model for individual values	Based on data from Amy Vickers (2002) Handbook of Water Use and Conservation

WMOST v1 User Manual and Case Study Examples

Description	Value Used in Model	Original Values	Data Source
Demand Management			
Price elasticity for each user	see case study model for individual values	see case study model for individual values	Based on Beecher 1994 - reviewed over 100 price elasticity of demand studies: residential: -0.2 to -0.4; industrial: -0.5 to -0.8
Capital cost to implement price increase	$23,000	$23,000	(Town of Breckenridge ~ 24,000 people served/day) Rogers, G. H. (2004). "Water Conservation Plan, Town of Breckenridge." Accessed April 20, 2005. http://www.townofbreckenridge.com/documents/page /Water%20Efficiency%20Plan%202004.pdf
O&M cost to administer price increase (e.g., resurvey for appropriate price etc.)	$2,000/yr	$2,000/yr	(Town of Breckenridge ~ 24,000 people served/day) Rogers, G. H. (2004). "Water Conservation Plan, Town of Breckenridge." Accessed April 20, 2005. http://www.townofbreckenridge.com/documents/page /Water%20Efficiency%20Plan%202004.pdf
Maximum price change over planning horizon	20%	20%	BPJ based on existing tiered pricing structure and demand management practices
Initial cost by providing rebates	$3,186,600	Total cost for high-efficiency appliances; Ideally, the user should determine the anticipated annual rate of use of rebates by households in order estimate annual cost rather than total maximum use and one initial cost.	Total Households in 1990=9382+1240=10622 Rebates $100/device-dishwasher, washing machine, high efficiency toilet- $300/household $300x10622=3,186,600
O&M cost of providing rebates	$0	None	BPJ
Maximum demand reduction	0.6 MGD	Total reduction in daily demand from high efficiency appliances	Washers = 8000 gal/hh/yr Dishwashers = 2150 gal/hh/yr Toilets = 1058 gal/hh/yr Total Households = 10622
SEPTIC			
Percent septic use for each user	9.4%	99% Danvers is sewered, ~2,155 on-site septics in Middleton (SWMI Ph1) * 2.68 ppl per HH = 64.3% of pop on septics (total pop is 8987). Middleton accounts for 14% of water. So 64% of 14% is 9.4% on septic.	MA DEP ASRs 2000-2012, Personal communication with Derek Fullerton/Director of Public Health/Middleton
SURFACE WATER			
Reservoir Storage			
Initial reservoir volume	533 MG	75% of active volume	BPJ
Minimum reservoir volume	0	0	Since active volume specified in report, assume that volume is max and min is zero (http://pubs.usgs.gov/sir/2006/5044/pdf/SIR2006-5044.pdf page 12)
Current maximum reservoir volume	710 MG	710 MG active volume	see above
Capital construction cost	$1,542,790/MG ($2013)	$1,542,790/MG ($2013)	Avg of EPA 2003 and Reading MA on finished water above ground storage
O&M costs	$15,428/MG ($2013)	1%	BPJ; did not see line item in town budgets so set at a minimum value
Streamflow			
Inflow from external surface water	cfs; See model for inidividual values	Sum of RIV_FLOW from Reaches: 16, 17, 23, 32	HSPF RIV_FLOW parameter
In-stream flow standards	cfs; See model for monthly values	Seasonal streamflow criteria for FL3 applied to August median unaffected flow adjusted for difference between WMOST and HSPF flows	SWMI Phase 1 Report
Maximum flow standard	NA	NA	None known
Private withdrawals of surface water	0	Not available	None known
Private discharge of surface water	0	Not available	None known

Description	Value Used in Model	Original Values	Data Source
GROUNDWATER			
Groundwater recession coefficient	0.01	1 minus area weighted average of AGWRC for HRUs in study area	HSPF
Initial groundwater volume	1134 MG	Sum AGWS across all HRUs for Day 1 of simulation (HSPF output in inches, mult by blended areas IMPL & PERL to get volume)	HSPF parameter AGWS
Minimum volume	706 MG	Sum AGWS across all HRUs (HSPF output in inches, mult by blended areas IMPL & PERL to get volume), take min MGD over simulation period and add 10%	HSPF parameter AGWS
Maximum volume	2838 MG	Sum AGWS across all HRUs (HSPF output in inches, mult by blended areas IMPL & PERL to get volume), take max MGD over simulation period and add 10%	HSPF parameter AGWS
Flow from external groundwater	0	0	No upper subbasins with groundwater draining to study area
Private withdrawals of groundwater	0	Not available	None known
Private discharge of groundwater	0	Not available	None known
INTERBASIN TRANSFER			
Purchase price for IBT potable water	$3,803/MG	$3,032/MG+$771/MG	MWRA water cost (http://www.mwra.state.ma.us/finance/intro.htm): $3,032/MG plus cost for operating distribution system and administrative functions $771 (based on Middleton water budget divided by MG since Middleton only buys from Danvers and distributes)
Purchase price for IBT wastewater	$6,340/ MG	$5,930,789/935 MG	Based on Danvers wastewater division annual budget divided by estimated 2012 MG wastewater flow
Initial cost for new/additional IBT potable water	$29,500,000/MGD	Water demand is 0.27 mgd (4 07-3.72) 3.72 mgd is WMA authorized withdrawal volume. 4.07 is 20-yr demand from DCR MWRA connection cost: $8 mil for 0.27 MGD inclusive of joining fee and construction costs	http://www.wickedlocal.com/weston/news/x868522107/At-MWRA-water-use-drops-but-expenses-dont $5 mil/mg to join About $250/lf to build water line (5 miles) ($5 mil/mgd*.27mgd)+$250/lf*5miles*5280ft/mile=$7.95 $7.95 mil/0.27mgd=$29,500,000/MGD
Initial cost for new/additional IBT wastewater	0	Already exists	Assume additional flow would not require capital contribution
Daily limit for wastewater	6 MGD	6 MGD	Based on estimated existing use of SESD
INFRASTRUCTURE			
Planning horizon	20 years	20 years	SWMI permitting horizon
Interest rate	5%	5%	EPA Community Water System Survey 2000

WMOST v1 User Manual and Case Study Examples

Water Treatment Plant

Description	Value Used in Model	Original Values	Data Source
Customer's price for potable water	$5.03/HCF	Residential: $5.03/hcf (0-20 hcf), 5.60 (20-24), 7 26 (over 24 hcf); Base fee per quarter: $10.50/HCF --> base fee was not included, unknown number of connections	MA DEP RGPCD Ave: 56.2, Ave Danvers HH = 2.42 ppl, Ave HCF/HH/mo = 5.45 HCF (1st tier), Town of Danvers website: http://www.danvers.govoffice.com/index.asp?Type=B_BASIC&SEC=%7BC6D1F088 6D0B-470A-A159-5734F9D4C585%7D
Gw pumping – Capital construction cost	$747,285/MGD ($2013)	$747285/MGD ($2013)	EPA Water need survey 2003, CCI data to update costs
Gw pumping –O&M costs	$0/MG	0	Based on town budgets, included in WTP O&M
Gw pumping –Current max capacity	1.74 MGD	Max daily over timeperiod +10% = 1.74 MGD	MA DEP ASRs 2000-2012
Gw pumping lifetime -remaining on existing construction	33 years	33 years	Assume same as WTP, no other data source identified
Gw Pumping lifetime- new construction	35 years	35 years	Assume same as WTP, no other data source identified
Sw pumping – Capital construction cost	$453,885/MGD ($2013)	$453,885/MGD ($2013)	EPA Water need survey 2003, CCI data to update costs
Sw pumping –O&M costs	$0/MG	0	Based on town budgets, included in WTP O&M
Sw pumping –Current max capacity	10.42 MGD	Max daily over timeperiod +10% = 10.42 MGD	MA DEP ASRs 2000-2012
Sw pumping lifetime -remaining on existing construction	33 years	33 years	Assume same as WTP, no other data source identified
Sw Pumping lifetime- new construction	35 years	35 years	Assume same as WTP, no other data source identified
Wtp - Capital construction cost	$2,022,884/MGD	$17.7 million in 2011$, CCI 2011 to 2013 = 1.0743	CCI data, and: http://www.wickedlocal.com/danvers/news/x1852609581/Danvers-water-treatment-plant-receives-17-million-bid#axzz2Os0ZSAmf
Wtp -O&M costs	$5,314/MG	$5,754,928/ 1083 MG	Based on town budgets' water division annual budget divided by 2012 MG demand
Wtp lifetime -remaining on existing construction	33 years	Built in 1976, bid for renovations in 2011, 24 mo construction (33 years)	http://www.wickedlocal.com/danvers/news/x1852609581/Danvers-water-treatment-plant-receives-17-million-bid#axzz2Os0ZSAmf
Wtp lifetime- new construction	35 years	35 years	http://www.wickedlocal.com/danvers/news/x1852609581/Danvers-water-treatment-plant-receives-17-million-bid#axzz2Os0ZSAmf
Wtp-Current max capacity	9.4 MGD	9.4 MGD	MA DEP ASRs 2000-2012
Capital cost of survey & repair	$774,368	$774,368	Based on MWRA project in Lynnfield, MA scaled to miles of pipe in Danvers and Middleton (0.62% of pipes need fixing, $145/ft (in$2004) detection and repair)
O&M costs for continued leak repair	$77437/yr	10% of capital	BPJ
Maximum percent of leaks that can be fixed	99%	99%	BPJ

Description	Value Used in Model	Original Values	Data Source
Water reuse facility			
Capital construction cost	$10,402,467/MGD	$6 million/ MGD ($1996)	Asano 1998; since small towns assume plant size between 1-5 MGD; assume upgrade from secondary to tertiary treatment that meets nonpotable reuse and ASR standards
O&M costs	$2,850/MG	10% of capital, convert to MG	BPJ
Lifetime remaining on existing construction	0 yrs	No existing WRF	
Lifetime of new construction	35 years	35 years	BPJ
Current maximum capacity	0 MGD	No existing WRF	
Nonpotable water distribution system			
Consumer cost for nonpotable water	$3.02/HCF	60% of potable price	http://www.irwd.com/customer-care/understanding-your-bill/recycled-water-rates.html
Capital construction cost for nonpotable distribution system	$12,529,440/MGD	$80,188,416 for all pipes; assume no more than half of demand met by nonpotable	EPA 2003 average cost per foot of distribution main: $93.16 (2003$) = 135.60 ($2013). SWMI Phase 1 - 112 miles of pipe; Max Np use is ~6.4 MGD
O&M cost for nonpotable distribution system	$1,716/MG	5%	Assume since new pipes, less O&M than usual 10%
Aquifer Storage and Recovery			
Capital construction cost	$10,807,824 /MGD		EPA Water need survey 2003 for injection well costs and Asano 1998 for treatment cost to meet gw recharge standards
O&M costs	$3,769/MG		EPA Water need survey 2003 for injection well costs and Asano 1998 for treatment cost to meet gw recharge standards
Lifetime remaining on existing construction	0 yrs	No existing ASR	
Lifetime of new construction	35 years	35 years	BPJ
Current maximum capacity	0 MGD	No existing ASR	
MEASURED FLOW			
Measured flow	cfs; See model for inidividual values	Sum of RIV_FLOW from Reaches: 37 and 46	HSPF RIV_FLOW parameter

United States
Environmental Protection
Agency

Office of Research and Development
National Health and Environmental
 Effects Research Laboratory
Atlantic Ecology Division
Narragansett, RI 02882

Official Business
Penalty for Private Use
$300

PRESORTED STANDARD
POSTAGE & FEES PAID
EPA
PERMIT NO. G-35